Advanced Introduction to International Water Law

Elgar Advanced Introductions are stimulating and thoughtful introductions to major fields in the social sciences, business and law, expertly written by the world's leading scholars. Designed to be accessible yet rigorous, they offer concise and lucid surveys of the substantive and policy issues associated with discrete subject areas.

The aims of the series are two-fold: to pinpoint essential principles of a particular field, and to offer insights that stimulate critical thinking. By distilling the vast and often technical corpus of information on the subject into a concise and meaningful form, the books serve as accessible introductions for undergraduate and graduate students coming to the subject for the first time. Importantly, they also develop well-informed, nuanced critiques of the field that will challenge and extend the understanding of advanced students, scholars and policy-makers.

For a full list of titles in the series please see the back of the book. This is also available on https://www.elgaronline.com/ and https://www.advancedintros.com/ for Elgar Advanced Introduction in Law.

Advanced Introduction to
International Water Law

OWEN MCINTYRE

*School of Law & Environmental Research Institute,
University College Cork, Ireland*

Elgar Advanced Introductions

Cheltenham, UK • Northampton, MA, USA

© Owen McIntyre 2023

All rights reserved. No part of this publication may be reproduced, stored in a retrieval system or transmitted in any form or by any means, electronic, mechanical or photocopying, recording, or otherwise without the prior permission of the publisher.

Published by
Edward Elgar Publishing Limited
The Lypiatts
15 Lansdown Road
Cheltenham
Glos GL50 2JA
UK

Edward Elgar Publishing, Inc.
William Pratt House
9 Dewey Court
Northampton
Massachusetts 01060
USA

A catalogue record for this book
is available from the British Library

Library of Congress Control Number: 2023946858

This book is available electronically on Elgar Advanced Introductions: Law
www.advancedintros.com

ISBN 978 1 80220 670 8 (cased)
ISBN 978 1 80220 671 5 (eBook)
ISBN 978 1 80220 672 2 (paperback)

Printed and bound in Great Britain by
TJ Books Limited, Padstow, Cornwall

Contents

Preface		vi
1	The conceptual setting for international water law	1
2	History and evolution of international water law	13
3	Equitable and reasonable utilisation	31
4	Prevention of significant transboundary harm	70
5	Cooperation and procedural rules of international water law	90
6	Environmental protection and ecosystems conservation	110
7	Institutional arrangements for transboundary water resources management	134
8	International water law and legal convergence	153
9	International water law and the Sustainable Development Goals—SDG 6	171
Index		197

Preface

Unlike several of the other volumes in this excellent series to which I've had the good fortune to be able to refer, this book is in fact intended to serve as a reasonably comprehensive, entry-level introduction to the discrete and rapidly evolving sub-field of international water law, rather than a more ambitious critical reflection upon its current (and future) trajectory, which mainly covers new developments in practice or academic research. There are several reasons for opting for this approach. Firstly, despite a wealth of ground-breaking research recently published in this area, there doesn't exist such an entry-level introductory text purporting to cover all the key constituent elements of international water law. In addition to one highly authoritative and seminally important text covering the entire field, which could neither be described as short nor introductory, there exist various commentaries on key instruments, handbooks on the broader field of water governance, and a rich selection of edited collections covering specific important aspects of the field, such as inter-State dispute resolution. However, a gap exists among these offerings for an introduction that might provide an interested graduate student (of international law) with a broad overview of this sub-field and a sense of its growing systemic relevance and importance. Secondly, it is a relatively compact sub-field, even though it is increasingly thoroughly intertwined with a range of other sub-fields of international law, and so it is eminently feasible to aim to provide a reasonably comprehensive introductory account of international water law.

However, in addition to providing a reasonably basic introduction to the topic which is accessible to a reader with little in-depth knowledge of the specialist subject-matter, as an 'advanced' species of the genre it also seeks to engage with a range of the key developments and challenges emerging in the field which are proving (or are likely to prove) increasing impactful and formative for the normative approaches adopted. These include: the emergence and continuing elaboration of the so-called 'ecosystem approach' to transboundary water management, and this approach's quite profound legal implications; the phenomenon of 'convergence' by means of which international water law borrows from (and contributes to) the normative forms adopted and employed in other, related sub-fields of international law; and the emergence of the SDGs as a novel global governance paradigm for implementing the environmental and developmental goals identified as central to realisation of the universally accepted overarching framework of sustainable development. This volume also endeavours, therefore, to the extent possible in a work of this brevity, to map out the legal implications of each of these defining developments.

At the request of the publisher, the text seeks to minimise the use of references, limiting these to the most systemically foundational instruments or the most important academic contributions, thereby providing an informed starting point for further reading and research on many of the most critical aspects of international water law or other directly related sub-fields.

This volume therefore attempts, perhaps unwisely, to achieve a number of aims—to provide an entry-level, comprehensive introduction; to address several of the key developments and challenges impacting international water law; and to provide a useful 'jumping-off' point for scholars embarking on new research projects in this vibrant and dynamic area. If pursuit of this range of aims should prove impractical within the confines of a short, coherent and readable volume, the folly is entirely my own.

1. The conceptual setting for international water law

1.1 Introduction

Few would now deny that reliable access to adequate supplies of fresh water has become a pressing global problem. Estimates suggest that by 2030 global water requirements will nearly double those in 2005 and will exceed current reliable supply levels by 40 per cent. The resulting water stress that many States will suffer will alter established patterns of agricultural production, adversely affect natural ecosystems and the services they provide, and generally impact livelihoods and living standards. As the freshwater crisis has come to be recognised as one of the key environmental crises of the twenty-first century, the perception of water as a valuable natural resource has at last taken hold among policymakers. Of course, this is equally true of transboundary water resources, where competition between co-basin States for the right to use shared waters has intensified. Thus, the role of international water law in promoting equitable transboundary water resources management and effective hydro-diplomacy becomes ever more critical.

As a discrete body of rules and principles, international water law is a relative newcomer. However, States have long engaged in formal cooperative arrangements over the use of shared international rivers and lakes. This is evidenced by the fact that a Mesopotamian agreement on the utilisation of shared water resources provides us with the oldest known example of an international treaty, while the Rhine Commission provides the first example of a permanently constituted international (intergovernmental) organisation. However, in the modern practice of international law relating to shared water resources, States had until recently been principally concerned with rights of navigation upon international rivers and lakes, with the International Law Association's (ILA) 1966 adoption of the seminal Helsinki Rules providing the first codification of international rules applying to the utilisation and protection of shared water resources.[1] The first global conventional instrument on the use of shared international water resources, the UN Watercourses Convention, was not adopted by the UN General Assembly until 1997 and only obtained the 35 ratifications required to enter into force in August 2014.[2] Similarly, measures to amend the 1992 United Nations Economic Commission for Europe (UNECE) Water Convention[3] so as to open it up to global membership, only entered into effect in 2013. Despite the immense importance of groundwater resources for meeting human needs around the world, international law relating to the utilisation and protection of transboundary aquifers is an even more recent and less developed field. It was only in 2008 that the International Law Commission (ILC) adopted its non-binding Draft Articles on Transboundary Aquifers,[4] and in 2014 that the parties to the UNECE Water Convention adopted their Model Provisions on Transboundary Groundwaters, intended to support improved cooperation over shared groundwater resources.[5]

Though relatively recently elaborated, international water law is already quite well settled around three key rules: the principle of equitable and reasonable utilisation, widely regarded the cardinal rule in the field; the duty to prevent significant transboundary harm; and the duty to cooperate in the management of shared waters. The effective implementation of these rules infers a range of related, ancillary requirements, including various procedural rules that facilitate inter-State communication, and substantive rules that further inform the applicable due-diligence standards required of watercourse States. Since the initial identification and codification of prevailing State practice in the 1966 Helsinki Rules, almost all water resources agreements—global and regional framework conventions as well as river basin and boundary waters agreements—have

---

[1] Helsinki Rules on the Uses of the Waters of International Rivers, ILA, *Report of the Fifty-Second Conference* (Helsinki 1966) 484.
[2] UN Convention on the Law of the Non-Navigational Uses of International Wat-ercourses (adopted 21 May 1997, entered into force 17 August 2014), (1997) 36 *International Legal Materials* 700.
[3] 1992 UNECE Convention on the Protection of Transboundary Watercourses and International Lakes (adopted 17 March 1992, entered into force 6 October 1996) 1936 UNTS 269.
[4] UNGA Res. 63/124 (15 January 2009). ILC, *Report of the International Law Commission on the Work of its Sixtieth Session* (2008) UN Doc A/63/10.
[5] Model Provisions on Transboundary Groundwaters (2014) UN Doc ECE/MP.WAT/40.

established legal regimes based more or less on this model. Most notably, the UN Watercourses Convention, the UNECE Water Convention and the ILC Draft Articles on Transboundary Aquifers all adopt similar approaches based on these three basic rules.

Despite this trend towards the convergence of international water law around three basic, yet broad and flexible, principles, it is clear that this body of rules is continuously interacting with, and is increasingly being shaped by, other prolific, dynamic and pervasive fields of normativity, including international environmental law, international human rights law and international investment law. While the central relevance of international environmental law to transboundary water management has long been self-evident, the social protection values inherent to equitable and reasonable utilisation, including in particular the priority traditionally accorded to safeguarding 'vital human needs' related to shared water resources, have become very closely intertwined with the discourse on the human right to water ongoing in international human rights law. Of course, developments in international water law commonly arise in connection with major investment projects, often involving foreign private or public sector investors, having the potential to impact upon the environment of an international watercourse, upon another State's right to utilise the shared waters in question, or upon local people's access to adequate water resources or services. Thus, tensions may arise with normative frameworks established in the field of international economic law concerning the legal protection of foreign investors or compliance with the environmental and social safeguard policies of multilateral development banks or other international financial institutions.

While the complex interrelationship between international water law and other sub-fields of international law will be touched upon later, it is first necessary to outline the relatively rapid and dramatic functional and conceptual evolution of modern international water law—from a body of rules primarily concerned with the need to accommodate competing national economic interests based on territorial sovereignty, to one that is increasingly understood as requiring meaningful inter-State engagement in pursuit of a common interest, i.e., that of optimal utilisation of the benefits flowing from shared water resources in a manner that is equitable as well as environmentally and socially sustainable. This is, of course, characteristic of the recent transition, clearly evident in international environmental law, from an international law of coexistence to an international law of cooperation.

1.2 Conceptual evolution of international water law

Historically, States sharing river basins have tended to invoke one of a number of largely incompatible theoretical approaches to territorial sovereignty, primarily dependent on their position on the watercourse, on which to base their purported right to utilise shared water resources.

1.2.1 Absolute Territorial Sovereignty

Firstly, the notion of 'absolute territorial sovereignty', traditionally favoured by upstream States, asserts that a basin State may freely utilise waters within its territory without having any regard to the rights of other co-basin States. Having absolute sovereignty over water resources while they lie or flow within its territory, a State may abstract, pollute or otherwise utilise those waters to an unlimited extent, but has no right to demand continued flow or otherwise to assert its rights against any co-basin State. This approach was epitomised in the so-called 'Harmon Doctrine', named after the US Attorney-General who first articulated the principle in 1895 in the course of a dispute with Mexico over the Rio Grande. However, it has garnered little support in State practice, judicial or arbitral practice, or in the work of publicists, despite having been often invoked by upstream States in the early stages of disputes over transboundary waters. Indeed, even the United States did not assert this very strident position for long, as it proceeded to conclude bilateral treaties with Mexico in 1906[6] and Canada in 1909,[7] which are more consistent with the principle of equitable and reasonable utilisation.

---

[6] Convention between the United States and Mexico Concerning the Equitable Distribution of the Waters of the Rio Grande for Irrigation Purposes, (1906) 34 Stat. 2953; *Legislative Texts*, No 75, at 232.

[7] Treaty Between the United States and Great Britain Relating to Boundary Waters and Questions Arising Between the United States and Canada (adopted 11 January 1909, entered into force 5 May 2010) 79 UNTS 260.

1.2.2 Absolute Territorial Integrity

The second approach, traditionally favoured by States predominantly located downstream on shared basins, revolves around the notion of 'absolute territorial integrity' and would confer a right upon a lower basin State to demand the continuation of the full flow of waters of natural quality from an upper co-basin State, but confers no right to restrict or impair the natural flow of waters from its territory into that of a still lower co-basin State. This approach is the antithesis of the Harmon Doctrine and would, in effect, grant a right of veto to a downstream or contiguous co-basin State, as its prior consent would be required for any change in the regime of an international watercourse. Though this approach was endorsed in one soft-law instrument adopted in 1911,[8] States have consistently expressed grave reservations, and it was unequivocally rejected in the 1957 *Lac Lanoux Arbitration (Spain v France)*.[9] As in the case of 'absolute territorial sovereignty', little of the State practice or work of publicists cited in support of 'absolute territorial integrity' stands up to serious scrutiny, which suggests that both theoretical positions were invoked primarily as preliminary bargaining positions rather than as legal principles considered likely to assist in the resolution of disputes.

1.2.3 Limited Territorial Sovereignty

The third approach provides the theoretical basis for the cardinal principle of equitable and reasonable utilisation. 'Limited territorial sovereignty' entitles each co-basin State to an equitable and reasonable use of waters flowing through its territory. This principle may be understood as a compromise between the former two approaches because the sovereign rights of the upper basin State and the territorial integrity of the lower basin State are each restricted by recognition of the equal and correlative rights of the other. This approach is supported by the notion that there exists a community of interest among all co-basin States, requiring an equitable balancing of State interests that accommodates the needs and uses of each State. To permit flexibility, the concept of 'equitable and reasonable' is deliberately vague and can only be determined in each

---

[8] International Law Institute, International Regulation regarding the Use of International Watercourses for Purposes other than Navigation—Declaration of Madrid (20 April 1911).
[9] (Award on 16 November 1957) (1961) 24 *International Law Reports* 101.

individual case in the light of all relevant factors including, ever more prominently, human needs and environmental protection. It is beyond debate, however, that this approach represents the prevailing theory of international watercourse rights and obligations today, largely because it has its doctrinal origins in the sovereign equality of States, under which all States sharing an international watercourse have equivalent rights to the use of its waters and/or to other related benefits. However, it is not possible to apply the principle to the resolution of abstract disputes. Its application to the resolution of actual disputes will depend on the particular circumstances of each dispute and watercourse, thus emphasising the distributive character of the equitable apportionment of benefits envisaged thereunder, and permitting the principle to play a pivotal role in the development of customary international water law as regards social and environmental protection. It provides for the establishment of factors that are to be taken into account in determining equitable and reasonable water utilisation and allocation of benefits deriving from the resource, and thus it provides a theoretical framework within which consideration of social and environmental protection can easily be conducted.

The principle of equitable and reasonable utilisation is very widely supported in State practice and international treaty law, judicial and arbitral practice, significant codifications and the work of publicists. However, the inherent complexity of the process of balancing diverse interests by weighing up various factors that might be relevant in basins that differ greatly in terms of their economic, social and environmental conditions means that widespread agreement is somewhat less forthcoming regarding how the principle might be applied in practice. Therefore, the use of institutional management arrangements including, in particular, permanent international river commissions or river basin organisations, is increasingly regarded as essential for its effective implementation, by providing frameworks within which a mutually acceptable formulation of the principle of equitable and reasonable utilisation might be agreed, and subsequently applied within a particular basin regime. For example, through an inter-State dialogue facilitated by a river basin organisation, basin States might agree on the relative value to be afforded to key factors, such as environmental protection of the shared watercourse or the imperative of economic development.

1.2.4 Common Management and Community of Interest

The fourth approach to international relations over transboundary water resources, that of 'common management', is therefore entirely compatible with the third and might be regarded as conducive, if not essential, to the ongoing development and elaboration of the principle of equitable utilisation. Under this approach, the drainage basin is regarded as an integrated whole and is managed to some extent as an economic, social and environmental unit, with the right to utilise water resources either vested collectively in the community of riparian States or divided among co-basin States by agreement, accompanied by the establishment of some form of inter-State institutional machinery to formulate and implement common policies for the management and development of the basin. Common management is an approach to managing shared water resources rather than a normative principle of international law, and as such it has been widely endorsed by the international community and by international bodies concerned with codification of this field of international law. While the UNECE Water Convention imposes a general obligation upon State parties to participate in the establishment of 'joint bodies',[10] the UN Watercourses Convention actively encourages watercourse States to enter into common management arrangements consistent with the requirements of the principle of equitable and reasonable utilisation, the duty to prevent significant transboundary harm and the general duty to cooperate.[11] In fact, such cooperation is so centrally important that one might well argue that a State's failure to participate actively in the procedural requirements inherent to equitable and reasonable utilisation, including in appropriate institutional cooperation, would suggest that that State's planned or actual use may not be equitable or reasonable under customary or conventional requirements.

The common management approach arises from the notion that there exists a 'community of interest' among co-basin States in an international watercourse, a doctrine that suggests that the interests of each State must be identified and safeguarded on the basis of equity, and that has received the consistent endorsement of international courts and tribunals, with the

---

[10] *Supra*, n. 3, art 9.
[11] *Supra*, n. 2, arts 8, 21, 24, 27, 28 and 33.

Permanent Court of International Justice (PCIJ) originally concluding in 1929 that:

> This community of interest in a navigable river becomes the basis of a common legal right, the essential features of which are the perfect equality of all riparian States in the use of the whole course of the river and the exclusion of any preferential privilege of any one riparian State in relation to the others.[12]

It is generally understood that the community of interest concept functions to inform concrete obligations of riparian States, in particular that of equitable and reasonable utilisation.

1.3 Challenges and opportunities in international water law

1.3.1 Fragmentation and Convergence

The broad corpus of rules, principles and objectives comprising international water law has grown exponentially in the decades since the 1966 adoption of the Helsinki Rules, due to the adoption of several hundred agreements and other instruments, ranging from highly focused bilateral arrangements through to global framework conventions, and further supplemented by a wealth of soft law declarations, resolutions and guidance. However, serious questions persist regarding the maturity of this body of rules and whether it provides a coherent basis for addressing the looming global water crisis. To a certain extent international water law may be considered a victim of its own success, at least in terms of the elaboration of treaties, where there exists a clear risk of treaty proliferation, congestion and fragmentation. The relevant international legal landscape might be considered congested, with applicable bilateral, regional and global agreements often addressing different issues individually (e.g., shared groundwater resources) and using inconsistent principles and instruments to address related problems. Such fragmentation could give

---

[12] *Case Relating to the Territorial Jurisdiction of the International Commission of the River Oder* (Judgment No 16) (1929) *PCIJ Reports*, Series A No 23, at 27. See also, *Case Concerning the Gabčíkovo-Nagymaros Project (Hungary/Slovakia)* (1997) *ICJ Reports* 7, para 85; *Case Concerning Pulp Mills on the River Uruguay (Argentina v Uruguay)* (Judgment) [2010] *ICJ Reports* 14, para 281.

rise to a range of difficulties extending beyond normative incoherence and inconsistency. For example, fragmentation between international water law and various international environmental regimes, (addressing, inter alia, biodiversity, wetlands, chemical pollution and climate change), can lead to overlapping responsibilities and inefficiency in the pursuit of environmental goals. Equally, fragmentation between international water law and rule complexes beyond international environmental and natural resources law, such as the international human rights, trade and investment regimes, threatens to undermine normative coherence and the effective application of international rules.

Counteracting the threat of fragmentation, however, one can observe the phenomenon of 'convergence' occurring in the ongoing elaboration of the broader international law framework, whereby different environmental and natural resources law sub-fields appear increasingly to borrow normative approaches from each other, and from beyond international environmental and natural resources law, in an ongoing process of normative interpenetration and cross-fertilisation. This process may be expected to produce positive benefits in terms of environmental outcomes, such as the potential role in environmental enforcement of mechanisms established beyond international environmental law, such as the World Trade Organisation (WTO) dispute settlement system's increasing readiness to give due consideration to the legitimate environmental commitments of States, or the growing concern of human rights courts and monitoring bodies with the environmental or water-related dimensions of international human rights. Such interactions between different regimes may be understood as a process of cognitive learning, by means of which related legal regimes learn from each other's experience. The phenomenon of convergence represents a critical point in the developmental maturity of international environmental and natural resources law, as a process of normative consolidation which largely addresses concerns over fragmentation and treaty congestion, and thereby sustains the unitary and systemic character of international law and bolsters its coherent application, often across multiple environmental sectors and media, thereby reflecting the broad scale and interconnectedness of the natural world. Therefore, despite the relative dynamism and growing specialisation of international water law as a specific sub-field of international law, it does not appear to be retreating from the generalised parameters of international law into an autonomous and compartmentalised regime.

1.3.2 Sustainable Development Goals

It is increasingly apparent that Sustainable Development Goal (SDG) 6 and the specific targets and related indicators set out thereunder, are likely to act as a strong catalyst for the continuing development of international rules for sustainable water management. The inclusive participatory process employed for the elaboration and adoption of the SDGs, as well as the institutional and administrative mechanisms emerging for their implementation and for monitoring progress in that regard, enhance the normative legitimacy of the core values enshrined in socially progressive water-related legal instruments, thereby conferring upon the SDGs the potential to transform the interpretation and continuing development of international water law regimes. For example, the emphasis in SDG Target 6.B on ensuring the participation of local communities in water management should do much to promote the (as yet) nascent elaboration of more inclusive and participatory procedural and institutional arrangements for transboundary water cooperation. Similarly, the universal commitments regarding inter-State cooperation over transboundary waters reflected in SDG indicator 6.5.2 provide a novel and unprecedented incentive for watercourse States to elaborate and establish permanent institutional frameworks for meaningful engagement.

SDG 6 can, therefore, impact profoundly on international water law in a range of ways which cohere with existing trends in its continuing evolution. For example, it is likely to impact the continuing discourse in national and international law on the human right(s) of access to water and sanitation and, thereby, the normative interpretation and application of the related concept of 'vital human needs' commonly employed in international river basin and water resources agreements. SDG 6 also strongly supports a greater focus on ecosystem-based approaches to water management and regulation and promotes participatory water governance at both the national and international levels. At the purely international level, SDG 6 does much to support and encourage more intensive transboundary water cooperation amongst watercourse States, particularly by means of the dedicated benchmark for transboundary cooperative arrangements provided in indicator 6.5.2.

SDG 6 embodies the solemn commitment of the international community to work towards ensuring, by 2030 at the latest, universal availability of safe and adequate water and sanitation services, along with sustainable management of the water resources on which such services directly depend. As such, it is quite clear that international, transnational and domestic legal frameworks will play a central role in shaping the actions necessary for the realisation of SDG 6, despite the latter's non-legally binding character. It is equally apparent, however, that the values set out under SDG 6 must inevitably exert significant influence upon the continuing evolution, elaboration and implementation of legal measures, principles and approaches related to water and sanitation services provision and to the sustainable management of water resources more generally. This is no less true in respect of international legal frameworks applying to shared transboundary water resources. There exists, therefore, a close, two-way inter-relationship between international law and the SDGs, spanning the mutually supportive approaches embodied in international water, environmental and human rights law and in SDG 6. These bodies of law continue to develop rules and principles intended to promote normatively broad and inclusive rights to water and sanitation and to require environmental protection of shared international water resources and the ecosystems dependent thereon. Of course, one could not reasonably expect the targets adopted under SDG 6 and the indicators developed subsequently to be capable of capturing and fully reflecting the complexity and increasing sophistication of the corresponding legal frameworks. For example, the discourse that has been ongoing now for over 20 years regarding the emerging right(s) to water and sanitation has identified normative elements, both substantive and procedural, that could not possibly be captured under a mere three indicators.[13] Similarly, the emergence of the so-called 'ecosystem approach' in international water law has given rise to a host of increasingly sophisticated technical standards and methodological approaches, such as those relating to environmental flows or to ecosystem services and arrangements for payment therefore, which are not even touched upon by indicator 6.6.1. However, the articulation and solemn adoption of SDG 6 by almost the entire international community of States represents a universal formal political commitment to such values, which can only serve further to legitimise and inform such emerging norms. In addition, the development and agreement of a comprehensive set of targets and indicators related to SDG 6 does much to promote a common understanding of water-related entitlements and duties, and to provide clear benchmarks for the elaboration and implementation of relevant rules.

---

[13] SDG 6.1.1, SDG 6.2.1a and SDG 6.2.1b.

1.4 Conclusion

Though the conceptual foundations of international water law are firmly rooted in traditional theories of territorial sovereignty, increasing recognition of the unique economic, social and environmental importance of water as a vital resource is likely to amplify the role of distributive conceptions of equity in addressing water-related entitlements under international law. While we are as yet only facing into an unprecedented global water crisis, driven by ever increasing water demand, growing problems of water pollution and the spectre of climate change, this evolution as regards the purpose of international law can already be detected in the manner in which values inherent to international human rights law and international environmental law permeate the development and application of international water law, emphasising, inter alia, the priority of meeting vital human needs and maintaining aquatic ecosystems and the services provided thereby. This process is likely to be intensified by the adoption and implementation of the SDGs, and in particular SDG 6, which will provide the backdrop for international water law's continuing evolution.

2. History and evolution of international water law

2.1 Introduction

This chapter aims to trace the discernible shift in the normative focus of international water law, from facilitating free navigation in large transboundary watercourse systems to equitable allocation and optimisation of rights to use shared water resources for economic purposes and thence to current concerns regarding the protection of watercourse ecosystems and maintenance of the essential ecosystem services provided thereby. Though this evolution inevitably has a profound impact on the normative content of international water law and on the cooperative institutional mechanisms required to give it meaningful effect, it has kept faith throughout with established legal principles and duties, i.e., the principle of equitable and reasonable utilisation, the duty to prevent significant transboundary harm, and the duty to cooperate in good faith.

The current 'environmental era' suggests that equitable and reasonable use of shared waters will increasingly have regard to environmental and ecological factors. Similarly, adverse impacts upon watercourse ecosystems will figure more prominently as significant transboundary harm that watercourse States should take reasonable measures to prevent. Likewise, cooperation will increasingly occur through mechanisms for inter-State procedural engagement stressing avoidance or mitigation of ecological harms, and employing, for example, environmental impact assessment of relevant projects and/or joint basin planning based upon an ecosystem approach. Of course, freshwater resources are also now linked closely to the climate system, reflecting greater awareness of river basins as complex ecosystems providing essential ecosystem services, and the emergence of sophisticated methodologies for identification, evaluation, and maintenance of ecosystem services and the equitable allocation of benefits deriving therefrom.

2.2 Historical evolution of international water law

While conflicts over water go back to earliest human history, the means for moderating such conflicts and supporting cooperation are also age-old. Over the millennia, and especially in recent centuries, the form and content of such arrangements have evolved.

2.2.1 An Age-Old Concern

The need to manage conflict and cooperation over water resources has been apparent to law-makers from the very earliest times, as evidenced by the Babylonian Code of Hammurabi, which dates back almost 4000 years and was centrally concerned with the water use rights and duties of neighbouring landowners. Safe water management and its efficient allocation amongst users provided a basis for the emergence of bureaucratic societies, giving rise to the great fluvial civilizations of antiquity. As protection, development, and apportionment of freshwater resources have long been central societal concerns, the cooperative, or at least peaceful, management of shared freshwaters has been one of the earliest concerns of early diplomacy, even predating the nation State. A water-sharing agreement concluded around 3100 B.C. between the Mesopotamian city States of Umma and Lagash provides the earliest known treaty. Historical records identify fifteen agreements relating to eight different international river basins that were concluded before the modern system of nation States was created by the 1648 Peace of Westphalia.

Individual agreements comprising the Westphalia peace settlement expressly provided for freedom of navigation on transboundary watercourses.[1] Thereafter, diplomatic engagement concerning shared waters increased rapidly, with a further 13 agreements concluded by the end of the 1600s and 81 agreements concluded during the 1700s. Commentators note that this accelerating pace of treaty activity was due to the dissolution of certain European empires, which 'internationalised' additional watercourses, and, even more importantly, changes in the ways in which humans used fresh water. The key historical change in water use was a shift towards increased non-navigational uses, which were often con-

---

[1] Article 9 of the Osnabrück Agreement concluded between Sweden, France, and Germany (1648).

sumptive in nature, or which at least impacted upon the quality of water resources remaining and, thus, upon the options for their future use.

2.2.2 From Navigation to Utilisation

While there are medieval agreements concerning such matters as territorial boundaries[2] and international legal claims regarding non-navigational uses of shared waters dating back to the mid-nineteenth century,[3] navigational issues predominated until relatively recently. A study of over 2000 international water agreements concluded by 2000 CE, found that the percentage dealing principally with navigation 'peaked in the period 1700–1930', while agreements focused on 'allocation and use issues were most significant as a percentage of total agreements negotiated during the period 1931–2000'.[4] This twentieth-century shift in focus from navigation to utilisation is reflected in the 1919 Treaty of Versailles,[5] which contained, in addition to provisions on freedom of navigation, provisions on such non-navigational uses as hydropower, irrigation, fishing, and water supply. Rivers provided important avenues for transport and trade, especially on large river systems traversing continental land masses and providing land-locked States with access to the sea and to wider maritime trade, and so freedom of navigation emerged as a general practice and principle of international law. Whereas the 1648 Peace of Westphalia provided for free navigation along the lower Rhine, a 1792 French Decree, concerning the *Opening of the Scheldt River to Navigation*, declared that free navigation along the entire course of an international watercourse was the inalienable right of all riparian States.[6] This position was quickly adopted in a number of instruments concerned primarily with rights of

---

[2] Treaty between Sweden (Finland) the Principality of Novgorod (Russia) concerning Lake Ladoga (1312); Treaty between the Bishop of Constance and Switzerland concerning Lake Constance (1554).

[3] A claim by the Netherlands in 1865 is generally understood to be the first diplomatic assertion of any such rule of international law. See, S.C. McCaffrey, *The Law of International Watercourses* (2nd ed.) (Oxford University Press, Oxford, 2007), at 62.

[4] E. Brown Weiss, 'The Evolution of International Water Law', (2009) 331 *Recueil des Cours* 163.

[5] Treaty of Peace between the Allied and Associated Powers and Germany (signed 28 June 1919) 225 *Consolidated Treaty Series* 189.

[6] Décret du 16 Novembre (1792).

navigation in international rivers.[7] The Treaty of Versailles declared parts of the rivers Rhine, Moselle, Meuse, Elbe, Oder, and Danube to be 'international' and created a suite of international commissions charged with securing freedom of navigation on each, including freedom of navigation rights for non-riparian States, a position supported judicially.[8]

The rapid industrialisation of Europe and North America from the mid-nineteenth century, and the related growth of urban settlements and of irrigated agriculture, led to unprecedented increases in freshwater consumption. For example, the United States saw a nearly a three-fold increase in water use between 1900 and 1930, with irrigated land in the western US almost doubling in area during this period. Industrialisation also led to significant increases in water use for hydropower generation, mineral extraction and processing, and myriad manufacturing processes, often facilitated by the construction of large-scale water-related infrastructure, made possible by means of newly developed engineering techniques. Such dramatic changes in States' water use inevitably impacted upon the subject-matter of inter-State water cooperation[9] and disputes.[10]

Codification of international rules applicable to non-navigational uses of shared watercourses became a priority for learned associations, leading to the 1961 adoption by the Institut de Droit International of the Salzburg Resolution on the Use of International Non-Maritime Waters[11] and,

---

[7] Treaty of Peace and Alliance between the French and Batavian Republic, art 18, (16 May 1795), concerning the Rhine, the Meuse, the Scheldt, and the Hondt (6 *Martens* 532); Principal Resolution of the Imperial Deputation (Reichsdeputationshauptschluss) (25 February 1803), concerning the portion of the Rhine shared between the Bavarian and the Swiss Republic (3 *Martens Supp* 239); Treaty Demarcating the Frontiers between Prussia and Westphalia (signed 14 May 1811) arts 7, and 9. L.A. Teclaff, 'Fiat or Custom: The Chequered Development of International Water Law', (1991) 31 *Natural Resources Journal* 45, at 47–48.

[8] *Case Relating to the Territorial Jurisdiction of the International Commission of the River Oder* (Judgment No 16) (1929) *PCIJ Reports*, Series A No 23.

[9] League of Nations, *Report of the Commission of Enquiry to the Barcelona Conference* (Geneva 1921).

[10] *Lac Lanoux Arbitration (Spain v France)* (Award on 16 November 1957) (1961) 24 *International Law Reports* 101.

[11] Institut de droit international, 'Utilisation of Non-Maritime International Waters (Except for Navigation)', (Salzburg 1961) 40 *Annuaire de l'Institut de Droit International* 381 ('Salzburg Resolution').

most notably, the International Law Association's 1966 Helsinki Rules on the Uses of the Waters of International Rivers.[12] This latter soft-law instrument established the template for subsequent agreements addressing non-navigational uses. In addition to a chapter on 'Navigation', the Helsinki Rules contained chapters on 'Equitable Utilisation of the Waters of an International Drainage Basin', on 'Pollution', and on 'Procedures for the Prevention and Settlement of Disputes'.

The International Law Association was primarily concerned with the emerging problem of competing 'economic' uses of international watercourses that grew more rapidly in the nineteenth and early-twentieth centuries than the applicable legal frameworks. As with most subsequent instruments,[13] the Helsinki Rules formulate the principle of equitable and reasonable utilisation as the legal entitlement of each basin State 'to a reasonable and equitable share in the beneficial uses of the waters of an international drainage basin', whilst providing an open-ended list of relevant factors to be considered in the determination of such a share, having due regard to the circumstances in each case. These factors are understood as creating a framework for reconciling the respective interests of watercourse States, based on a distributive conception of equity emphasising each State's economic and social dependence upon the water resources in question. Although the Helsinki Rules address the problem of pollution of an international drainage basin, they only require prevention or abatement of pollution which would cause substantial injury in the territory of a co-basin State. In 1966, the International Law Association was only concerned with pollution of an international watercourse to the extent that it impacted on economic or social uses of the shared waters by another basin State. The Helsinki Rules also identify a suite of procedural rules to assist basin States in the prevention and settlement of disputes regarding shared waters. Most notably, 'with a view to preventing disputes from arising between States as to their legal rights or other interests', basin States 'should in particular furnish to any other basin State, the interests of which may be substantially affected, notice of any proposed

---

[12] International Law Association, 'Final Report of the Helsinki Rules', *Report of the Forty-Seventh Conference of the International Law Association* (Helsinki 1966) 484 ('Helsinki Rules').
[13] Most notably, the UN Convention on the Law of Non-Navigational Uses of International Watercourses (21 May 1997, entered into force 17 August 2014) UN Doc No A/51/869, arts 5 and 6.

construction or installation which would alter the regime of the basin'. Once again, balancing the economic and social interests of basin States was the primary concern.

In 1970 the UN General Assembly recommended that the International Law Commission 'take up the study of the law of the non-navigational uses of international watercourses with a view to its progressive development and codification',[14] and shortly thereafter the Commission commenced two decades of analysis of relevant State, treaty, judicial, and arbitral practice, culminating in the adoption in 1994 of its Draft Articles on the Law of the Non-Navigational Uses of International Watercourses.[15] On this basis, the Sixth (Legal) Committee of the General Assembly negotiated the final text of the UN Watercourses Convention, which was adopted on 21 May 1997, and eventually entered into force in August 2014 upon receiving its 35th ratification. The Convention was inspired and largely shaped by the International Law Association's earlier codification of customary international water law in the Helsinki Rules, and includes a broadly similar formulation of the cardinal principle of equitable and reasonable utilisation, a broader articulation of the subordinate duty to prevent significant transboundary harm, either due to pollution or otherwise, and a clear statement of the general obligation of States to cooperate 'in order to attain optimal utilisation and adequate protection of an international watercourse'. The obligation to cooperate is supplemented by detailed procedural rules, which focus particularly on the duty of inter-State notification concerning planned measures with possible adverse effects. Again the balancing of economic and social interests and uses remained the key concern throughout the Convention's extended elaboration.

The Convention's value as an authoritative statement of customary international law,[16] and as a template for subsequent regional or water-

---

[14] *Progressive Development and Codification of the Rules of International Law Relating to International Watercourses*, G.A. Res. 2669, UN GAOR, Twenty-Fifth Sess., Supp. No 28, (8 December 1970) UN Doc A/8028, at 127.
[15] International Law Commission, 'Draft Articles on the Law of Non-Navigational Uses of International Watercourses', *Report of the Forty-Sixth Meeting of the International Law Commission* (May 2–July 22, 1994) UN Doc A/49/10, at 195.
[16] See *Case Concerning the Gabčíkovo-Nagymaros Project (Hungary/Slovakia)* (1997) *ICJ Reports* 7, para 85.

course agreements,[17] stems largely from the fact that the International Law Commission spent 20 years developing the text, a process that involved, inter alia, a detailed questionnaire addressed to UN Member States on the topic and a total of 15 reports prepared by eminent jurists acting as Special Rapporteurs. Customary international law continues to evolve, and developments subsequent to the drafting of the Convention text might require its reinterpretation in certain respects, particularly as regards the requirements of environmental and ecological protection.[18] The most recent comprehensive study of how customary international law has evolved has produced the International Law Association's Berlin Rules on Water Resources,[19] a 2004 revision and updating of the seminal Helsinki Rules which draws heavily from more recent developments in international environmental and human rights law.

2.3 The current state of affairs: the imperative of ecological sustainability

Pollution control has long been centrally important in international water law,[20] with agreements dealing principally with environmental pollution and the protection of ecosystems mainly emerging after 1950. Few modern watercourse agreements do not include detailed provisions for environmental protection of the watercourse and its related ecosystems. Even where watercourse instruments predate environmental concern and lack such provisions, they may be amended to incorporate such values[21] or interpreted by international tribunals as requiring such protection, especially in the light of subsequent environmental commitments of the

---

[17] See Protocol on Shared Watercourse Systems in the Southern African Development Community (SADC) Region, (2001) 40 *International Legal Materials* 321, art 2(2), (6), (7), (7 August 2000).
[18] See *Pulp Mills on the River Uruguay (Argentina v Uruguay)* (Judgment) [2010] *ICJ Reports* 14, para 204.
[19] ILA, *Report of the Seventy-First Conference of the International Law Association* (2004) 334 ('Berlin Rules').
[20] See *Lac Lanoux Arbitration, supra*, n. 10, 101.
[21] Minute 319 (20 November 2012), amending Treaty Respecting Utilisation of Waters of the Colorado and Tijuana Rivers and of the Rio Grande (3 February 1944) 3 UNTS 314.

States concerned.[22] Implementation of the environmental obligations inherent in international water law is aided by the parallel elaboration of corresponding rules under international environmental law instruments, and their considerable normative specificity. The central importance of environmental values is confirmed by the fact that every inter-State water dispute leading to judicial resolution in recent years has concerned environmental matters.[23] Of course, environmental provisions in international water instruments reflect corresponding concern in the domestic practice of water law and policy.

Though the environmental requirements of international water law are firmly established, they continue to evolve and recent years have seen a shift in emphasis from pollution prevention towards more wide-ranging protection of riverine ecosystems, reflecting growing awareness of watercourses as complex and fragile ecosystems providing a range of indispensable ecosystem services and requiring holistic management of a variety of interconnected ecological elements. This has obvious implications for the normative scope of international water law, requiring 'the adoption of less economic-oriented criteria for the management of freshwater resources, following a so-called "ecosystem approach"' and requiring consideration of the whole system rather than individual components. Before conclusion of the UN Watercourses Convention, a trend towards broader ecosystem obligations in international water law was discernible in both basin agreements[24] and regional framework conventions.[25] The continuing normative development of the ecosystem approach to the management of shared water resources will be critically important for realisation of the key emerging imperative of modern international water law, i.e., the optimal and ecologically sustainable use of shared water

---

[22] *Kishenganga Arbitration (Pakistan v India)* (Partial Award on 18 February 2013) (Permanent Court of Arbitration), para 452; *Gabčíkovo-Nagymaros* case, *supra*, n. 16, paras 112, 140.

[23] *Gabčíkovo-Nagymaros* case, *supra*, n. 16; *Pulp Mills* case, *supra*, n. 18; *Kishenganga Arbitration, supra*, n. 22; *Certain Activities Carried Out by Nicaragua in the Border Area (Costa Rica v Nicaragua)* and *Construction of a Road Along the San Juan River (Nicaragua v Costa Rica)* (Judgment on 16 December 2015) (2015) *ICJ Reports* 665.

[24] Great Lakes Water Quality Agreement (22 November 1978) 30 UST 1383, arts I and II; Agreement on Cooperation for Sustainable Development of the Mekong River Basin (5 April 1995) 2069 UNTS 3, arts 3 and 7.

[25] Protocol on Shared Watercourse Systems in the Southern African Development Community (SADC) Region (28 August 1995), art 2(3), (11), (12).

resources in an era of looming freshwater scarcity.[26] International water law will have to contend with inexorably rising demand for water, food, and energy, and associated large-scale water resources utilisation, as well as the very significant hydrological and ecological challenges posed by climate disruption and the mitigation and adaptation measures required to address it.

The UN Watercourses Convention goes beyond rules for mere management of pollution to reflect greater scientific understanding and concern regarding maintenance of watercourse ecosystems. Requiring watercourse States to 'utilise an international watercourse in an equitable and reasonable manner', the Convention stresses that they do so 'with a view to attaining optimal and sustainable utilisation thereof and benefits therefrom ... consistent with adequate protection of the watercourse'.[27] The factors identified therein as relevant have also evolved to include 'ecological and other factors of a natural character' and '[c]onservation, protection, development and economy of use of the water resources of the watercourse'.[28] The duty to prevent significant transboundary harm undoubtedly encompasses ecological disturbance in the light of heightened environmental sensibilities and the wealth of guidance on ecosystems management produced under multilateral environmental agreements.

Most significantly, the UN Watercourses Convention includes an entire Part IV, containing a suite of provisions expressly addressing the protection, preservation and management of an international watercourse, including express obligations concerning protection of international watercourse ecosystems, prevention of significant harm from pollution, introduction of alien species, and protection of the marine environment. Part IV opens with Article 20 containing the general obligation to protect and preserve watercourse ecosystems. The International Law Commission defined a (watercourse) ecosystem as an 'ecological unit consisting of living and non-living components that are interdependent and function as a community.' Ecosystem protection appears to enjoy priority, with the related provisions each addressing an aspect of this overarching imperative. For example, Article 21, containing the

---

[26] See Berlin Rules on Water Resources, arts 7, 8, 22–35, 38, 40 and 41.
[27] UN Watercourses Convention, art 5.
[28] Ibid., art 6(1).

Convention's pollution control obligations, defines 'pollution of an international watercourse' broadly to include 'any detrimental alteration in the composition or quality of the waters of an international watercourse which results directly or indirectly from human conduct'. Thus, Article 21 may also apply, in addition to more conventional cases of pollution, to uses that decrease watercourse flow resulting in damage to ecosystems. The International Law Commission emphasises the customary status of the obligation to preserve and protect watercourse ecosystems, declaring that '[t]here is ample precedent for the obligation ... in the practice of States and the work of international organizations'. It also links this obligation to the overarching principle of equitable and reasonable utilisation, explaining that it 'is a specific application of the requirement contained in Article 5 that watercourse States are to use and develop an international watercourse in a manner that is consistent with adequate protection thereof'. The Commission justifies the priority afforded to ecosystems protection by explaining that 'protection and preservation of aquatic ecosystems help to ensure their continued viability as life support systems, thus providing an essential basis for sustainable development'.

The only other globally applicable water resources convention, the UNECE Water Convention includes extensive and detailed provisions on the conservation and restoration of the ecosystems of shared basins,[29] which have inspired similar provisions in subsequently adopted basin agreements.[30] Early guidelines adopted under the UNECE Convention elaborate upon the implications of the ecosystem approach.[31] The rules on international groundwater resources are evolving similarly, with the International Law Commission's Draft Articles on Transboundary Aquifers stressing 'the role of the aquifer or aquifer system in the related

---

[29] UNECE Convention on the Protection and Use of Transboundary Watercourses and International Lakes ('UNECE Water Convention') (17 March 1992, entered into force 6 October 1996) 1936 UNTS 269, arts 1(2), 2(2)(b), (d), 3(10(i).

[30] Convention on the Protection of the Rhine ('Rhine Convention') (22 January 1998, entered into force 1 January 2003), arts 2, 3, 5; Convention on Cooperation for the Protection and Sustainable Use of the Danube River ('Danube Convention') (29 June 1994, entered into force 1998), arts 1(c), 2(3), 2(5); Framework Agreement on the Sava River Basin (3 December 2002, entered into force 29 December 2004) 2367 UNTS 688, art 11(a).

[31] UNECE, *Guidelines on the Ecosystem Approach in Water Management* (December 1993) UN Doc ECE/ENVWA/31.

ecosystem' and requiring States 'to ensure that the quantity and quality of water retained in an aquifer or aquifer system, as well as that discharged through its discharge zones, are sufficient to protect and preserve such ecosystems'.[32] On the other hand, the ILA's 2004 Berlin Rules on Water Resources include a Chapter V, which is rich in ecological obligations, as well as Chapter VI on impact assessments, Chapter VII on responses to ecological extremes, and several relevant articles in Chapter VIII on groundwater.

The continuing evolution of the international law applicable to water resources includes the obligation to conserve ecosystems as set out in the Convention on Biological Diversity (CBD).[33] The CBD Conference of the Parties (COP) has adopted two successive programmes of work on inland water that clarify the Convention's application to watercourse ecosystems.[34] Relevant guidance has been developed under the CBD on a range of relevant issues, such as invasive alien species.[35] Lessons learned in implementing other environmental conventions concerned with water-related ecosystems, such as the Ramsar Convention,[36] inform ecosystems obligations arising in international water law. The rich technical guidance developed under the Ramsar regime clarifies such obligations. This complementarity reflects the critical role of wetlands in the functioning of aquatic ecosystems and the provision of important ecosystem services.

---

[32] International Law Commission, Draft Articles on Transboundary Aquifers, *Report of the International Law Commission on the Work of Its Sixtieth Session*, (2008) II *Yearbook of the International Law Commission* UN Doc A/CN.4/SER.A/2008/Add.1, at 19, arts 5.1(i), 10; also UNECE, Model Rules on Transboundary Groundwaters (2014), Provision 2.1.

[33] Convention on Biological Diversity ('Biodiversity Convention') (approved 22 May 1992, entered into force 29 December 1993) 1760 UNTS 79, arts 1, 8(f); also Biodiversity Convention Decision V/6, 'Ecosystem Approach' (Nairobi 22 June 2000) UN Doc UNEP/CBD/COP/5/23.

[34] Biodiversity Convention Decision IV/4 (1998), Annex I; Biodiversity Convention Decision VII/4 (2004), Annex.

[35] Biodiversity Convention Decision VIII/27 (2006).

[36] Convention on the Protection of Wetlands of International Importance, Especially as Waterfowl Habitat ('Ramsar Convention') (opened for signature 2 February 1971, entered into force 21 December 1975) 996 UNTS 245; Ramsar Conference of the Parties Resolution IX.1 (Kampala 2005), Annex A.

2.4 Future implications for international water law

International instruments stipulating an ecosystem approach to shared watercourse management introduce an unprecedented measure of technical complexity to the field of international water law, requiring detailed normative elaboration of hitherto opaque commitments of watercourse States regarding ecosystems protection. While consensus is emerging regarding core elements of this obligation, it is increasingly apparent that an ecosystem approach will require more sophisticated rules on procedural engagement regarding water resources, including rules to facilitate meaningful stakeholder participation, along with robust institutional structures to facilitate complex benefit-sharing arrangements.

In addition to pollution prevention, ecosystem obligations include a range of measures, including the maintenance of environmental flows, support for ecosystem services, arrangements for more adaptive management of water resources, and enhanced procedural rules and institutional arrangements.

2.4.1 Environmental Flows

The International Union for the Conservation of Nature (IUCN) has defined environmental flows as 'the water regime provided within a river, wetland, or coastal zone to maintain ecosystems and their benefits where there are competing uses and where flows are regulated', the overarching goal of which 'is to provide a flow regime that is adequate in terms of quantity, quality, and timing for sustaining the health of the rivers and river systems', while also stressing the significance of social and economic factors. A regime of environmental flows is central to adoption of an ecosystem approach to international watercourse management, which is increasingly understood as essential for achieving riverine ecological health, sustainable development, and the sharing of benefits between users. Though seldom addressed expressly or directly in international water instruments, the legal character of environmental flow requirements must be understood as a key element in taking an ecosystem approach.

A requirement to maintain minimum environmental flows, derived from established principles of international environmental law, was supported in a Permanent Court of Arbitration finding that 'hydro-electric projects

... must be planned, built and operated with environmental sustainability [and minimum environmental flow in particular] in mind'.[37] Similarly, the ICJ has recognised the legal significance of maintaining flows for ecological purposes.[38] Numerous studies support the emergence of a related legal obligation and, as the legal nature of the obligation to maintain flows becomes clearer, through the accumulated practice of international courts, water convention secretariats, and national regulators, the science is similarly advancing. Assorted actors, including environmental convention secretariats, civil society organisations, and academic researchers have produced technical and methodological guidance on effective implementation of environmental flow requirements.

2.4.2 Ecosystem Services

With publication of the 2005 Millennium Ecosystem Assessment, which introduced the concept to mainstream transboundary water governance, the maintenance of 'ecosystem services' quickly emerged as the overarching objective of an ecosystem approach, and thus of any environmental flow regime. It facilitates assessment of the beneficial services provided by natural ecosystems that provides a methodology for their valuation and consideration within decision-making processes. It presents the prospect of improved inter-State cooperation and thus of optimised and equitable use and more effective protection of shared watercourse ecosystems. In this role, the ecosystem services concept can function, in theory at least, to facilitate avoidance and resolution of inter-State water disputes.

Despite their relative novelty, relevant methodologies are increasingly employed in transboundary water cooperation. Guidance on water resources management for the maintenance of ecosystem services has been developed under the Ramsar Convention and the Convention on Biological Diversity,[39] which has even adopted targets relating to the ecosystem services provided by inland waters.[40] An emerging legal obligation to maintain ecosystem services is supported, inter alia, by recent statements of the UN Special Rapporteur on Human Rights and the

---

[37] *Kishenganga Arbitration, supra,* n. 22, paras 450–452, 454, and (Final Award), 20 December 2013.
[38] *Certain Activities* case, *supra,* n. 23, para 105.
[39] Biodiversity Convention Decision VII/4 (2004), Annex.
[40] Biodiversity Convention Decision X/2 (2010), Annex, para 13.

Environment framing the issue as a human right of access to ecosystem services.[41]

An ecosystem services regime may include arrangements for payment for ecosystem services and, though not extensively developed in international legal practice, key actors in the field provide guidance on how such payment systems might work.[42] Payments might assist inter-State engagement over transboundary waters where such arrangements may be utilised as one element of broader integrated benefit-sharing arrangements and may provide a useful means of equitably rebalancing competing State interests in a shared watercourse potentially giving rise to water disputes. It is clear that in most cases benefit-sharing would require some element of redistribution or compensation tailored to the specific circumstances.

2.4.3 Adaptive Management

'Adaptive management', involving a strategy that is 'flexible, iterative and responsive to the constantly changing conditions of both complex ecosystem processes and available scientific knowledge',[43] is increasingly regarded as central to the effective application of an ecosystem approach.[44] It addresses fundamental uncertainty regarding the functioning of complex dynamic socio-ecological systems, the value of certain ecosystems and their services, and the potential effects of certain policies and projects on the functioning of ecosystems. Such uncertainty is exacerbated, and adaptive strategies ever more necessary, with the threat of climate disruption to freshwater ecosystems. Adaptive management seeks to ensure ecosystem 'resilience' by adopting a systematic approach to adapting and improving natural resources management by learning from previous management interventions. Incorporation of adaptive measures into conventional legal frameworks is problematic due to traditional prioritisation of stability over flexibility in legal regimes, especially where such regimes function to facilitate investment in large-scale water

---

[41] *Report of the Special Rapporteur on the Issue of Human Rights Obligations Relating to the Enjoyment of a Safe, Clean, Healthy and Sustainable Environment* (19 January 2017) UN Doc A/HRC34/49, at 4.
[42] UNECE, *Guidance on Water and Adaptation to Climate Change* (2009).
[43] V. De Lucia, 'Competing Narratives and Complex Genealogies: The Ecosystem Approach in International Environmental Law', (2015) 27 *Journal of Environmental Law* 91–117, at 93.
[44] Biodiversity Convention Decision V/6.

infrastructure and the avoidance or resolution of water disputes. Robust joint institutional mechanisms will be necessary to implement adaptive management.

The procedural rules of international water law are most highly developed regarding major development or water-use projects, where conventional instruments provide for inter-State notification and, where necessary, structured consultation and negotiation.[45] Traditionally, outcomes of inter-State procedural engagement regarding such water use or infrastructure projects have understandably emphasised legal stability, as evidenced by consistent judicial recognition of the role of environment impact assessment (EIA) in such engagement.[46] An EIA generally comprises a one-time, 'front-loaded' process that attempts to predict and mitigate adverse impacts in advance of the commencement of a project. An adaptive regime would accommodate uncertainty through flexible decision-making procedures, allowing 'incremental and gradual changes that transition experimentally to new standards or arrangements, while monitoring, assessing and adjusting these changes and their effects'. Therefore, legal frameworks for transboundary cooperation must evolve to create suitable institutions employing highly sophisticated procedures for inter-State engagement.

Strong links exist between adaptive management and the precautionary principle, as both seek to accommodate scientific uncertainty, and the former can be regarded as a means of implementing the latter, which enjoys extensive support as customary law. Consistent ICJ endorsement of a requirement for 'continuing' environmental assessment might amount to tacit judicial recognition of the importance of adaptive ecosystem-based management in certain situations of scientific uncertainty.[47]

---

[45] UN Watercourses Convention, arts 11–19.
[46] *Pulp Mills* case, *supra*, n. 18, para 204.
[47] *Gabčíkovo-Nagymaros* case, *supra*, n. 16, Separate Opinion of Judge Weeramantry, paras 108–110; *Nuclear Tests case (New Zealand v France)* (1995) (Request for an Examination of the Situation in Accordance with Paragraph 63 of the Court, Judgment of 20 December 1974) *ICJ Reports* 344; Legality of the Use by a State of Nuclear Weapons in Armed Conflict (1996) *ICJ Reports* 140; *Pulp Mills* case, *supra*, n. 18, para 205.

2.4.4 Procedural Rules and Institutional Arrangements

The UN Watercourses Convention elaborates extensively on procedural rules to moderate the risk of conflict.[48] The Berlin Rules largely mirror the procedural requirements[49] of the UN Watercourses Convention, but go somewhat further, drawing on the regime of international human rights law to recognise participatory rights of the public and stakeholders and providing guidance on the necessary institutional arrangements.[50]

2.4.4.1 *Stakeholder and public participation*

By largely neglecting to provide for meaningful participation of stakeholders or the wider public in the management of shared transboundary water resources, international water law appears somewhat out of step with developments in general international law. Water agreements typically focus on inter-State engagement to the exclusion of public participation, as epitomised by Part III of the UN Watercourses Convention. Though many functioning common management institutions do in fact seek actively to engage with a broader range of stakeholders, including local communities, civil society and research institutions, a formal legal framework for such engagement tends to be lacking. Similarly, though the conduct of an EIA of major projects in international watercourses is understood to be a 'requirement under general international law',[51] and though the national EIA frameworks to be applied would ordinarily entail extensive public disclosure, consultation and participation, this cannot coherently address the lacuna in international water law.

Accepting that effective public or stakeholder participation is crucial for the protection of transboundary watercourse ecosystems, and thus for achieving optimal and sustainable utilisation, it follows that new rules on procedural engagement are required which reflect the participatory rights emerging in the related fields of human rights law and environmental

---

[48] UN Watercourses Convention, arts 11–19.
[49] Berlin Rules on Water Resources, arts 56–63.
[50] *Ibid.*, arts 11 (duty to cooperate), 64–67 (joint arrangements or institutions).
[51] *Pulp Mills* case, *supra*, n. 18, para 204.

law.[52] This will require a corresponding development of the mandates and capacity of basin institutions.

2.4.4.2 Benefit-sharing arrangements

Greater focus on ecosystem services and related environmental flow obligations presages more extensive reliance upon 'benefit-sharing' arrangements in order to optimise beneficial use of ever-scarcer water resources while maintaining watercourse ecosystem integrity. Such arrangements might typically involve some form of payments for benefits, or for costs associated with enhanced stewardship of a transboundary watercourse, normally taken on by an upstream State. Ecosystem services provide a methodology for economic and social valuation of the benefits of watercourse ecosystems, including non-marketable benefits, which allows their integration into benefit-sharing arrangements. Such methodologies, where they are widely accepted, can help to provide a common starting point for benefit-sharing negotiations.

Crafting and implementing complex benefit-sharing arrangements will require intensive, highly technical, and continuing inter-State engagement, likely to prove beyond the capacity of current procedural rules and institutional structures. The experience of the US and Canada on the ground-breaking agreement concerning the Colombia river[53] suggests that, due to the inherent complexity of the considerations and calculations involved, a sophisticated legal and institutional framework for cooperation was required for formulating viable proposals. The relative success of different benefit-sharing regimes will depend upon effective procedural requirements and competent institutional arrangements for water cooperation and engagement with all stakeholders. Benefit-sharing arrangements focused on preservation of watercourse ecosystems and maintenance of ecosystem services will require highly capacitated, perma-

---

[52] Convention on Accession to Information, Public Participation in Decision-Making and Access to Justice in Environmental Matters ('Aarhus Convention') (25 June 1998, entered into force 30 October 2001) 2161 UNTS 447; Regional Agreement on Access to Information, Public Participation and Justice in Environmental Matters in Latin America and the Caribbean ('Escazú Agreement'), (9 April 2018, entered into force) 22 April 2021.

[53] Treaty Relating to Cooperative Development of the Water Resources of the Columbia River Basin, (17 January 1961) 15 UST 1555. Renegotiation of the Treaty commenced in May 2018.

nent institutions capable of facilitating intensive, inclusive, and ongoing engagement.

2.4.4.3 International water institutions

One might reasonably expect that joint institutions for cooperative transboundary water resources management will continue to proliferate and to develop in terms of their functional responsibilities. International water engagement has long been moving towards cooperative or joint management regimes, which again impacts upon the normative content of international water law by creating ever greater reliance upon the elaboration of sub-conventional rules, procedures and guidance by river basin organisations (RBOs) set up for the purpose. One might argue that this reflects the relative maturity of international water law, in terms of widespread acceptance of its key tenets and awareness of the impending water crisis. The trend towards the establishment of permanent RBOs is likely to receive a significant boost from global implementation of SDG 6, including target 6.5 which promotes transboundary cooperation and indicator 6.5.2 which suggests that watercourse States should focus on the role of formal institutional machinery established to facilitate meaningful cooperative engagement over transboundary waters.

2.5 Conclusion

As a sub-field of international law, international water law has evolved, firstly from a body of rules focused upon navigation in international watercourses to one primarily concerned with economically important non-navigation uses of shared waters, and, yet again, to a legal framework largely concerned with the conservation of riverine ecosystems and maintenance of the services provided thereby. In the case of each such transition, international water law is merely responding to changing patterns of water use and to changing requirements for water resources protection. The latest, ongoing transition is a response to current and imminent challenges, including the looming water, biodiversity and climate crises. Of course, this shift in the purpose and focus of international water law will inevitably have significant and wide-ranging implications for the normative and institutional structure of this evolving sub-field.

3. Equitable and reasonable utilisation

3.1 Introduction

The principle of equitable and reasonable utilisation represents a necessary compromise between two extreme and uncompromising legal perspectives regarding the rights conferred upon States, by virtue of their territorial sovereignty, to utilise shared transboundary water resources found within or passing through their territory. The first position, based on the theory of 'absolute territorial sovereignty' and traditionally favoured by upstream States, supports the argument that a co-basin State may freely utilise waters within its territory without having any regard to the rights of downstream or contiguous States. Having absolute sovereignty over water resources while these are present within its territory, a State may utilise or alter the quality or quantity of these waters to an unlimited extent, but accordingly has no right to demand continued flow or quality from another co-basin State. This approach is closely associated with the so-called 'Harmon Doctrine', named after the US Attorney-General who first elaborated the principle in the context of a dispute with Mexico over the waters of the Rio Grande. Harmon did not recognise any general legal requirement for the US to safeguard the supply of water to Mexico, stating that the question of whether the US should 'take any action from considerations of comity' was one which 'should be decided as one of policy only, because, in my opinion, the rules principles and precedents of international law impose no liability or obligation upon the United States'.

The second position, based on the theory of 'absolute territorial integrity' and traditionally invoked by downstream States, would confer a right on a co-basin State to demand the continuation of the full flow of waters of natural quality from another (upper or contiguous) co-basin State, but confers no right to restrict or impair the natural flow of waters from its territory into that of any other (still lower) co-basin State. This approach is the antithesis of the Harmon Doctrine and would effectively grant a right of veto upon a downstream or contiguous State, as its prior consent would be required for any change in the regime of the international watercourse. However, on the basis of detailed analysis of State practice regarding these two extreme contrasting positions, one can conclude that they were principally invoked as tools of advocacy rather than legal principles considered likely to assist in the resolution of actual inter-State disputes. It is worth noting, for example, that shortly after Attorney-General Harmon made his (in)famous statement of opinion in 1895, the US concluded bilateral treaties with Mexico (in 1906) and Canada (in 1909) which are very much more consistent with the principle of equitable and reasonable utilisation.

Over time practically all States sharing transboundary basins have come to adopt a third approach, based on the theory of 'limited territorial sovereignty' which represents a common-sense compromise between the absolutist positions outlined above. This approach recognises that both the sovereign utilisation of one (often upstream) basin State and the right to territorial integrity of another (often downstream) basin State are each restricted by a recognition of the equal and correlative rights of the other State. This approach is usually articulated in legal terms as the principle of 'equitable and reasonable utilisation', which entitles each co-basin State to an equitable and reasonable use of transboundary waters flowing through its territory, or to an equitable and reasonable share of the benefits deriving therefrom. As it involves recognition of the 'equality of right' of both upstream and downstream States, or of States causing and suffering pollution, equitable and reasonable utilisation coheres with the notion of the sovereign equality of States, a fundamental principle of public international law authoritatively enshrined in Article 2(1) of the UN Charter. However, equality of right does not entitle each State to an equal share in the waters of a shared basin, but only to an equal right *vis-à-vis* its co-riparian neighbours to an equitable share of the uses and benefits of the watercourse having regard to all relevant factors. The principle is also based on the notion that there exists a 'community of interest' among all co-basin States, requiring a fair balancing of State interests which accommodates the reasonable needs and uses of each State to the greatest extent possible. To permit the necessary flexibility in its application, the concept of 'equitable and reasonable' use is consciously understood as normatively vague, requiring that it be determined in each individual case in the light of all relevant factors including, most notably, the human, economic and social dependence of each State upon the water resources in question, as well as considerations of environmental protection. In essence, the principle of equitable and reasonable utilisation requires that, in using shared water resources, each co-basin State must have due regard for the legitimate needs and interests of other co-basin States. Clearly, this requires each basin State to proceed with a reasonably comprehensive objective appreciation of the relevant factual circumstances, suggesting that all basin States need to engage in meaningful procedural cooperation in order to meet their commitments under the principle. Unquestionably providing the prevailing normative framework for identifying international watercourse rights and obligations today, equitable and reasonable utilisation has its doctrinal origins in the sovereign equality of States, whereby all States sharing an international watercourse have equivalent rights to the use and benefit of its waters.

Though equitable and reasonable utilisation is universally regarded as the pre-eminent substantive rule of international water law, a normative framework requiring the equitable balancing of the legitimate water-related interests of basin States must inevitably involve intense inter-State procedural engagement, which often can only be facilitated by the establishment of technically competent inter-State institutional machinery. Such institutions can ensure effective inter-State communication involving, inter alia, prior notification of planned projects potentially impacting upon the watercourse, routine exchange of information regarding the utilisation or condition of the shared waters, or related expressions of concern on the part of any basin State. The pivotal role of institutional mechanisms in giving effect to the principle of equitable and reasonable utilisation has long been recognised by the international community, with Recommendation 51of the *Action Plan for the Human Environment* adopted at the 1972 UN Conference on the Human Environment (UNCHE) at Stockholm calling for the 'creation of river basin commissions or other appropriate machinery for cooperation between interested States for water resources common to more than one jurisdiction' and setting down a number of basic principles by which the establishment of such bodies should be guided. Although such institutional structures can take numerous different forms and may have diverse remits, there are today well over 100 river basin organisations (RBOs) performing a very extensive range of coordination and joint management functions. Reliance on such institutional mechanisms to facilitate the inter-State cooperation necessary to achieve equitable and reasonable utilisation is often referred to as the 'common management' approach, which further underlines the existence of a community of interest among co-basin States.

3.2 Legal status

The principle of equitable and reasonable utilisation enjoys very considerable support in the judicial deliberations of international and federal courts and tribunals. For example, in the *Gabčíkovo-Nagymaros* case before the International Court of Justice, judge ad hoc Skubiszewski, in this dissenting opinion, referred to the 'cannon of an equitable and reasonable utilization' as an expression of 'general law'.[1] The principle receives almost universal support in treaty law, international codifications, declaratory soft-law instruments and the general practice of States, as well as in the writings of leading publicists. Indeed, having regard to all widely recognised indicators of the existence of customary rules, the International Law Commission concluded unequivocally 'that there is overwhelming support for the doctrine of equitable utilization as a general rule of law for the determination of the rights and obligations of States in this field'.[2]

However, such near universal acceptance by States is due in large part to its flexibility and normative indeterminacy, with the principle providing both a rather vague aspirational goal to guide transboundary water cooperation and the starting point for a process to investigate, identify and ultimately reconcile the needs, interests, entitlements and obligations of interdependent co-basin States. Indeed, expert commentators tend to highlight the principle's role as a discursive process, emphasising its procedural and institutional aspects. Thomas Franck notably describes it as a striking example of certain 'sophist rules' existing in international law, which have a 'multi-layered complexity', by virtue of which they enjoy a degree of elasticity, and 'usually require an effective, credible,

---

[1] *Case Concerning the Gabčíkovo-Nagymaros Project (Hungary/Slovakia)* (1997) *ICJ Reports* 7, at 235.
[2] International Law Commission, *Report of the International Law Commission on the Work of its Forty-Sixth Session*, UN GAOR, Forty-Ninth Sess., Supp. No 10, (1994) UN Doc A/49/10, at 222.

institutionalized, and legitimate interpreter of the rule's meaning in various instances', a process which he characterises as 'institutionalized multilateralization'.[3] Of course, certain other commentators tend to be more critical of the principle's normative indeterminacy.

3.3 Normative content

In essence, the principle of equitable and reasonable utilisation involves the allocation of rights in the uses and benefits of shared water resources on the basis of a distributive conception of equity, requiring actors to have due regard to all relevant factors. This suggests that uses and benefits will be shared in proportion to each basin State's needs, where such needs are estimated through consideration of those factors which are accepted by the States concerned as relevant to water allocation. Therefore, the factors considered relevant to understanding each State's dependence on the shared waters, and thus to the estimation of each State's equitable and reasonable allocation of uses and benefits, are absolutely central to the principle's application and codified or conventional formulations of the principle usually include an accompanying indicative list of such relevant factors. Such a list was first set out in Article V(2) of the International Law Association's seminally important 1966 Helsinki Rules.[4] Most significantly, Article 6(1) of the 1997 UN Watercourses Convention[5] now lists the following factors as relevant:

1. Geographic, hydrographic, hydrological, climatic, ecological and other factors of a natural character;
2. The social and economic needs of the watercourse states concerned;
3. The population dependent on the watercourse in each State;
4. The effects of the use or uses of the watercourses in one watercourse State on other watercourse States;

---

[3] T. Franck, *Fairness in International Law and Institutions* (Clarendon Press, Oxford, 1995), at 67, 75, 81–82 and 140.
[4] International Law Association, Helsinki Rules on the Uses of the Waters of International Rivers, ILA, *Report of the Fifty-Second Conference* (Helsinki 1966) 484.
[5] Convention on the Law of the Non-Navigational Uses of International Watercourses (adopted 21 May 1997, entered into force 17 August 2014), (1997) 36 *International Legal Materials* 700.

5. Existing and potential uses of the watercourse;
6. Conservation, protection, development and economy of use of the water resources of the watercourse and the costs of measures taken to that effect;
7. The availability of alternatives, of comparative value, to a particular planned or existing use.

Such a list is not intended to be exhaustive and a range of additional factors might be relevant in the particular circumstances of a particular basin, negotiation or dispute, such as any religious, cultural or local customary significance attached to the river in question or to its waters. Similarly, the conduct of the States concerned regarding a contested use or project might be relevant including, for example, excessive delay in raising objections.

Whilst all key instruments emphasise the lack of any hierarchy among the relevant factors, it is apparent from the practice of States that certain considerations will usually be accorded more significance than others. For example, while Article 6(3) of the UN Watercourses Convention provides that '[t]he weight to be given to each factor is to be determined by its importance in comparison with that of other relevant factors', Article 10(2) would appear to prioritise 'vital human needs', a key element in identifying the 'population dependent on the watercourse in each state' as a relevant factor under Article 6(2). Of course, this elevation of vital human needs corresponds with the increasingly prominent global discourse on the human right(s) to water and sanitation and so enhances the human rights dimension of international water law. In this regard, it is worth noting that a statement of understanding agreed at the time of the adoption of the UN Watercourses Convention advises that 'special attention is to be paid to providing sufficient water to sustain human life, including both drinking water and water required for production of food in order to prevent starvation'.[6]

Indeed, it appears from the practice of States in this field that what matters above all else is the dependence of each watercourse State upon the shared waters in question, in terms of either human, social or economic needs, and that the relevant factors listed above and elsewhere largely function

---

[6] UN General Assembly Working Group, *Report of the Working Group to the General Assembly* (1997) UN Doc A/C.6/51/SR.57, at 3.

to elucidate the true nature and extent of such dependence. For example, though the UN Watercourses Convention suggests that existing and potential uses of an international watercourse will in principle be considered equally, with the International Law Commission noting that 'neither is given priority' and that 'one or both factors may be relevant in a given case',[7] it is likely that existing uses will be favoured as they can more easily be scrutinised in terms of their human, social, economic or environmental benefits (or adverse impacts), while the difficulties inherent in reliably considering the beneficial character (or negative impacts) of future uses are manifest. Equally, the examination of certain factors, such as efforts at conservation and economy of use of water resources by a particular State and the availability to a State of alternatives to a planned or existing use of shared waters, primarily help to inform that State's true dependence upon the contested waters. Further, though 'natural' factors, including the geography and hydrology of the basin, are listed first under both the 1966 Helsinki Rules and the 1997 UN Watercourses Convention, there is general agreement among expert commentators that such factors are of only marginal significance as these do not relate directly to a State's dependence upon the shared waters and so could undermine the distributive nature of the equitable allocation envisaged under the principle of equitable and reasonable utilisation. Nor would significant reliance upon such factors support the notion of the sovereign equality of co-basin States. The distributive nature of equity as applied in the particular field of international water law is highlighted by the fact that relatively little significance is attributed to the physical characteristics of the drainage basin, such as the length of the course of the river situated within each basin State, the extent of the drainage basin area lying within the territory of each basin State, or each State's relative contribution of water to the flow of the river. To apply the principle of equitable and reasonable utilisation otherwise might lead to an increased share for those basin States already abundant in water resources. The situation in international water law contrasts with the application of equitable principles in maritime territorial delimitation of the continental shelf, where the emphasis has traditionally been placed on the physical extent of each coastal State's coastline. Consistent with the emergence of the human rights dimension in international water law, this particular conception of equity may

---

[7] ILC, *supra*, n. 2, at 233.

amount to a recognition in international law of humanity's unique and total dependence upon freshwater resources.

Although the principle of equitable and reasonable utilisation has its origins in inter-State arrangements for allocating co-basin States' quantum share of transboundary waters, it is now routinely concerned with environmental requirements and the environmental consequences of incompatible uses. Indeed, as it seeks to balance economic, social and environmental imperatives in the use of shared waters, equitable and reasonable utilisation can now be understood as a means of operationalising the more broadly relevant objective of sustainable development in the specific context of transboundary water resources. Thus, it should come as no surprise that environmental protection and sustainability requirements are inherent to authoritative modern formulations of the principle. Accordingly, Article 6(1)(a) of the UN Watercourses Convention refers to 'ecological' factors, Article 6(1)(d) to the 'effects of the use or uses ... on other watercourses states', and Article 6(1)(f) to 'conservation, protection ... and economy of use of the water resources of the watercourse'. This connection is even more pronounced in the 1992 UNECE Water Convention, under which the focus is squarely on environmental protection with the parties required, inter alia, to ensure 'ecologically sound and rational water management ...[and] conservation of water resources and environmental protection' and 'where necessary, restoration of ecosystems'.[8] The environmental aspects of the principle have tended to enjoy ever increasing emphasis in recent years. In fact, two seminally important cases brought before the International Court of Justice relating to environmental protection of transboundary rivers have expressly linked the principle of equitable and reasonable utilisation to the overarching objective of sustainable development.[9]

---

[8] 1992 UNECE Convention on the Protection of Transboundary Watercourses and International Lakes (17 March 1992, entered into force 6 October 1996) 1936 UNTS 269, art 2(2).

[9] *Gabčíkovo-Nagymaros, supra,* n. 1, and *Pulp Mills on the River Uruguay (Argentina v Uruguay)* (Judgment) [2010] *ICJ Reports* 14.

3.4 The concept of 'equity' in international water law

The rather vague language of 'equity' pervades the various bodies of rules of international law which apply to the management, protection, utilisation, allocation and supply of freshwater resources, including international environmental law, international human rights law and of course international water resources law. While this inevitably results in considerable uncertainty regarding the precise normative implications of such rules, a measure of normative vagueness permits the flexibility required to secure the participation of hesitant State actors in the incremental progressive elaboration of the kind of sophisticated legal regimes necessary for the cooperative and sustainable management of an increasingly scarce and contested resource. It should also be remembered that no two watercourse systems are remotely alike—hydrologically, ecologically, climatically, socially, economically, demographically or politically—and so a measure of flexibility is welcome in the applicable rules and standards in order to facilitate the taking account of relevant contextual factors. It is possible, however, notwithstanding the need for vagueness and flexibility, to divine certain core elements and values inherent to the concept of equity, as least insofar as it applies to shared water resources.

3.4.1 International Environmental Law

International environmental law has long relied on the concept of equity in high-profile declaratory and conventional instruments including, for example, Articles 3(1) and 4(2)(a) of the 1992 United Nations Framework Convention on Climate Change[10] and Articles 1 and 15(7) of the 1992 Convention on Biological Diversity,[11] both of which instruments have a role in informing normative frameworks for water resources management as regards climate change adaptation and the maintenance of aquatic and water-related ecosystems. As early as 1978 the United Nations Environment Programme facilitated adoption of a seminal set of draft principles to guide States in the environmental protection and cooperative utilisation of shared natural resources, including water

---

[10] United Nations Framework Convention on Climate Change, (1992) 31 *International Legal Materials* 851.
[11] United Nations Convention on Biological Diversity, (1992) 31 *International Legal Materials* 818.

resources, which declared equity to be the key principle for 'controlling, preventing, reducing or eliminating adverse environmental effects which may result from the utilization of such resources'.[12] More generally, the concept of inter-generational equity articulated in Principle 3 of the Rio Declaration[13] and the notion of intra-generational equity encapsulated in the principle of common but differentiated responsibilities set out under Rio Principle 7 illustrate the absolutely central importance of the concept of equity to the principle of sustainable development, the key organising principle underlying all norms of modern international environmental law. Thus, 'in many respects, UNCED was about equity' largely because 'in the absence of detailed rules, equity can provide a conveniently flexible means of leaving the extent of rights and obligations to be decided at a subsequent date'.[14] Through its articulation in Article 5(c) of the 1992 UNECE Water Convention, the principle of inter-generational equity has been expressly endorsed in one of only two global instruments applying to shared natural resources, suggesting that the utilisation of shared waters in a manner or at a rate inconsistent with their natural capacity to be replenished would prejudice the right of future generations to enjoy the use of such waters or of dependent ecosystems. The reference in the 1997 UN Watercourses Convention to the goal of sustainable utilisation appears specifically intended to mitigate against such utilisation. The principle of equitable and reasonable utilisation can be understood to be very closely linked to sustainable development in that it requires consideration of a range of factors related to sustainable development of the resource, thereby providing the legal framework for operationalising the concept. Clearly, such inter-linkage compounds the relevance of the principles of inter-generational and intra-generational equity to the application of international water law.

However, while both principles are concerned with achieving some form of equitable distribution of costs and benefits in the use of environmental

---

[12] UNEP, Draft principles on conduct in the field of the environment for guidance of States in the conservation and harmonious utilization of natural resources shared by two or more States (19 May 1978) UNEP Governing Council Decision 6/14.

[13] UN Conference on Environment and Development (UNCED), Rio Declaration on Environment and Development, (1992) 31 *International Legal Materials* 874, UN Doc A/CONF.151/26 (vol 1).

[14] P. Sands and J. Peel, *Principles of International Environmental Law* (3rd ed.) (CUP, Cambridge, 2012), at 213–214.

resources, they primarily serve to identify the parties to whom equitable considerations should apply, rather than to inform the normative meaning of the notion of equity as it applies in the field of international environmental law. Though 'many environmental treaties refer to or incorporate equity or equitable principles ... treaties rarely provide a working definition of equity'.[15] Incorporation of the rather nebulous concept of equity can even cloud attempts to codify and articulate core, foundational rules of international environmental law, such as the general duty of States to prevent significant transboundary harm. For example, Articles 9 and 10 of the International Law Commission's 2001 Draft Articles on the Prevention of Transboundary Harm from Hazardous Activities require States to seek 'acceptable solutions regarding measures to be adopted in order to prevent significant transboundary harm ... based on an equitable balance of interests'.[16] As with all leading conventional and declarative formulations of the international water law principle of equitable and reasonable utilisation, the 2001 ILC Draft Articles merely provide an indicative list of factors relevant to achieving such equitable solutions.

3.4.2 International Human Rights Law

The language of equity is equally ubiquitous, though no less uncertain, in the international discourse on the human right to water and sanitation. Notably, General Comment No. 15, the seminal document adopted in 2002 by the UN Committee on Economic, Social and Cultural Rights setting out the legal origins and normative content in international law of the purported human right to water, alludes to the concept in several contexts. For example, in relation to the need to ensure access to water resources for agriculture in order to realise the right to adequate food, paragraph 7 calls upon State parties to ensure 'that disadvantaged and marginalized farmers, including women farmers, have equitable access to water and water management systems'.[17] Thus, equity is here associated with the obligation of State parties to avoid any form of discrimination. In

---

[15] *Ibid.*, at 119.

[16] International Law Commission (ILC), Articles on the Prevention of Transboundary Harm from Hazardous Activities, *Report of the International Law Commission*, Fifty-Third Sess. (2001), UN Doc A/56/10, at 148–170.

[17] Committee on Economic. Social and Cultural Rights (CESCR), General Comment No 15, The Right to Water: Articles 11 and 12 of the International Covenant on Economic, Social and Cultural Rights (26 November 2002) UN Doc E/C.12/2002/1.

turn, paragraph 27, which concerns the obligation of States to ensure that water services are affordable, provides that:

> Any payment for water services has to be based on the principle of equity, ensuring that these services, whether privately or publicly provided, are affordable for all, including socially disadvantaged groups. Equity demands that poorer households should not be disproportionality burdened with water expenses as compared to richer households.

Here the Committee appears to employ a form of distributive equity based upon the concept of proportionality, or at least on the avoidance of disproportionate costs for the poor.

More recently, the targets set out under Sustainable Development Goal 6, which commits the international community to 'ensure availability and sustainable management of water and sanitation for all' and is very closely linked to realisation of the human rights to water and sanitation, place considerable reliance on the notion of equity. Recent methodological guidance produced by UN-Water to inform monitoring of the SDG targets on drinking water and sanitation, explains 'equitable access to safe and affordable drinking water for all' under target 6.1 and 'equitable sanitation and hygiene for all' under target 6.2 in terms of 'progressive reduction and elimination of inequalities between population sub-groups'.[18] Thus, it appears that the equitable values inherent to the obligations imposed under the purported human rights to water and sanitation merely require the progressive elimination of discriminatory practices regarding access to drinking water and adequate sanitation rather than an urgent redistribution of resources in order to eliminate all and any inequality of access to services.

3.4.3 International Water Resources Law

It is in the field of international water law, however, that equity plays an absolutely central role as the key normative principle informing application of the fundamental legal rule entitling each co-basin State to an equitable and reasonable use of transboundary waters flowing through its territory or of the benefits deriving therefrom. As explained above, rights

---

[18] UN-Water, *Methodological Note: Proposed Indicator Framework for Monitoring SDG Targets on Drinking Water, Sanitation, Hygiene and Wastewater* (WHO & UNICEF, Geneva, 2015).

to water use and benefits will be shared in proportion to each basin State's needs, where such needs are calculated through consideration of relevant factors, such as those listed under Article 6(1) of the UN Watercourses Convention. However, such a general articulation tells us little about the nature of equity as a source of rules impacting on the determination of watercourse States' right to utilise shared waters, or about the means by which equitable principles can be operationalised in the practice of international water law. In fact, equity has historically been invoked in relation to a number of different roles and it is helpful to identify and differentiate between several of the roles that equity might play in facilitating the fair sharing of benefits deriving from shared water resources and ensuring effective environmental protection of international watercourses and their dependent ecosystems. In doing so, there is little benefit in exploring the notional competence of the International Court of Justice (ICJ) under Article 38(2) of its Statute to apply, with the consent of the parties to a dispute, *equity ex aequo et bono*, which is generally understood as not referring to rules of law, primary or supplemental, but to the Court's capacity to settle disputes on the basis of conciliation. This application of 'equity' does not refer to considerations lying within the rules of law and thus does not form a component part of the corpus of rules and principles that constitute international law. As State parties to disputes are understandably reluctant to give any international tribunal such wide and unfettered discretion, neither the ICJ nor its predecessor, the Permanent Court of International Justice (PCIJ), has ever decided a case *ex aequo et bono*. In the 1929 *Free Zones* case, the PCIJ was careful to disregard any consideration of *equity ex aequo et bono* in the absence of the clear and explicit agreement of the parties.[19] When employing the concept of equity, the ICJ has consistently made a point of clarifying that it is referring to a role other than that of *equity ex aequo et bono*:

> Whatever the legal reasoning of a court of justice, its decision must by definition be just, and therefore in that sense equitable. Nevertheless, when mention is made of a court dispensing justice or declaring the law what is meant is that the decision finds its objective justification in considerations lying not outside but within the rules ... There is consequently no question in this case of any decision *ex aequo et bono*.[20]

---

[19] *Free Zones* case *(France v Switzerland)* (1929) *PCIJ Reports*, Series A, No 22, at 5.
[20] *North Sea Continental Shelf* cases *(Germany/Denmark/Netherlands)* (Judgment on 20 February 1969) (1969) *ICJ Reports* 50, at 3 and 47.

Exploring the principal ways in which equity can operate in the field of international water law involves characterising equity in a number of legal roles, including as a general principle of international law, as means of ensuring the procedural fairness of inter-State communication and engagement arrangements, and as a substantive rule of apportionment of water quantum, water uses or water-related benefits. Of course, these roles are not mutually exclusive, but merely reflect the primary manner in which equitable principles tend to be invoked and applied in respect of international freshwater resources.

3.4.3.1 Equity as a general principle of law

The place of equity among the rules of international law is commonly understood as that of a general principle of law. The drafters of the ICJ Statute considered the 'general principles of law recognized by civilized nations' as belonging among the sources of international law 'in virtue of their social foundation and rational character', while Article 38(1)(c) of the Statute accommodates the evolution of general legal principles as they are formed in national legal systems through the ongoing clarification of the central idea of justice and implementation of this idea into rules. In the *River Meuse* case, Judge Anzilotti said of the principle inherent in the Roman law maxim *inadimplenti non est adimplendum* (he who fails to fulfil his part of an agreement cannot enforce that bargain against the other party) that it is 'so just so equitable, so universally recognised, that it must be applied in international relations also'.[21] In the same case, Judge Hudson observed that under 'Article 38 of the Statute, if not independently of that Article, the Court has some freedom to consider principles of equity as part of the international law which it must apply'.[22]

While the reference to 'general principles of law' in Article 38(1)(c), may include the various 'principles' of international law commonly included in natural resources and environmental treaties and declarative instruments, such as the precautionary principle, the polluter-pays principle, and the principle of common but differentiated responsibility, it certainly permits the Court to apply widely employed principles of national law where there might otherwise arise lacunae among the established rules

---

[21] *Diversion of Water from the River Meuse (Netherlands v Belgium)* (1937) PCIJ Reports, Series A/B, No 77, at 50.
[22] *Ibid.*, at 77.

of international law. These would include general principles of 'natural justice' 'accepted by all nations *in foro domestico*', which could operate 'to avoid any possibility of a *non liquet* where there may be gaps in the law'.[23] The doctrines of abuse of rights and good faith are often cited as examples of such general principles.[24] However, the most prominent, and normatively rich, of such general principles of natural justice is that of 'equity', which plays a particularly significant role in the establishment, operation and application of the rules of international natural resources law and may be defined in this context as 'considerations of fairness, reasonableness, and policy often necessary for the sensible application of the more settled rules of law'.[25] As the concept of equity and particular equitable principles are to be found in many national legal systems, equity can be included among the corpus of norms that constitute applicable international law. That international tribunals may apply equitable principles without the express authorisation of the parties to an inter-State dispute was confirmed by Judge Hudson in the *River Meuse* case, where he stated that 'what are widely known as principles of equity have long been considered to constitute a part of international law, and as such they have often been applied by international tribunals'.[26]

Rather than borrow mechanically from domestic law, however, international tribunals have only 'invoked elements of legal reasoning and private law analogies', so that 'general principles derived by analogy from domestic law are only marginally useful in an environmental context'.[27] Indeed, the marginal utility of traditional domestic equitable principles is illustrated inadvertently by Judge Hudson who, in his separate opinion in the *River Meuse* case, cites several of the traditional equitable maxims found in Anglo-American jurisprudence which he regarded as potentially relevant to transboundary water resource disputes, including 'he who comes to equity must come with clean hands', 'he who seeks equity must do equity', and 'equality is equity'. The first of these maxims might apply to require that a party to a dispute seeking a remedy under international law ought to have acted in good faith and have discharged all relevant pro-

---

[23] P. Birnie, A. Boyle and C. Redgwell, *International Law and the Environment* (3rd ed.) (Oxford University Press, Oxford, 2009), at 26–27.
[24] *Free Zones* case, n. 19, at 167.
[25] I. Brownlie, *Principles of Public International Law* (4th ed.) (Oxford University Press, Oxford, 1990), at 26.
[26] *Supra*, n. 21, at 76–77.
[27] Birnie, Boyle and Redgwell, *supra*, n. 23, at 27.

cedural and substantive obligations. In turn, the second might mean that the State that exploits the shared resource first may not object when the neighbouring State begins to do so or, conversely, that the State that succeeds in preventing the exploitation of the shared resource by a neighbour may itself be estopped from exploiting the resource. This is essentially what occurred in the *River Meuse* case, where the Netherlands' complaint against Belgium's diversion of their shared River Meuse was dismissed largely because the Netherlands had itself earlier engaged in a similar diversion scheme. The third equitable maxim cited by Judge Hudson might suggest that there ought to be a proportionate distribution of benefits and burdens in the use of shared resources by promoting reliance upon objective criteria indicating the extent of each State's dependence. Other maxims which might prove relevant to the equitable and reasonable utilisation of shared water resources could include 'equity will not suffer a wrong to be without a remedy', which might influence the application of rules on State responsibility and liability. Similarly, the maxim 'equity imputes an obligation to fulfil an obligation' might influence the application of rules on the enforcement of conventional obligations.

It is necessary to express caution, however, regarding 'the difficulty of drawing equitable principles from national legal systems and applying them in the international system'[28] and it is immediately apparent that not all equitable maxims will apply in the context of international watercourse law. Accordingly, one should be wary of assuming their direct relevance. For example, the maxims 'where the equities are equal, the first in time shall prevail' and 'delay defeats equity' might appear to support the so-called doctrine of 'prior appropriation' which, in the case of international watercourses, is effectively contradicted by Article 6 of the UN Watercourses Convention, which gives equal weight in principle to existing and potential uses of share waters, and has not anyway enjoyed general support in State practice.[29] It has also been widely criticised as potentially wasteful, not conducive to the optimal economic development of the watercourse, and potentially environmentally damaging. Similarly, the maxim 'where there is equal equity, the law shall prevail'

---

[28] V. Lowe, 'The Role of Equity in International Law', (1992) 12 *Australian Yearbook of International Law* 54–81, at 80.

[29] X. Fuentes, 'The Criteria for the Equitable Utilization of International Rivers', (1996) 65 *British Yearbook of International Law* 337–412, at 365.

might incorrectly be assumed to suggest that the established status quo in transboundary water resources regimes should not generally be disturbed.

As a general aspirational ideal the notion of equity is also of limited utility. Although many supportive statements from judicial and other actors characterise equity as a 'direct emanation of the idea of justice' or as necessary 'considerations of fairness, reasonableness and policy', they provide only the vaguest guidance for legal decision-makers regarding relevant normative or social values. Such statements do little to help us to understand how equitable principles might be applied by international lawyers in the settlement of international disputes over shared water resources. International courts have employed equitable principles to resolve natural resource disputes on a number of occasions, particularly in respect of maritime boundaries and resources. However, these cases have not elaborated greatly on how equity applies as a general principle of law. Rather, the ICJ has taken greater care to clarify precisely what roles equity as a general principle of law cannot play. Thus, one might conclude that such judicial deliberations 'amount to no more than a bundle of highly impressionistic ideas' and, further, that when 'employed in this way "equitable principles" become highly faint indications of the reasoning ... on which judicial discretion has been exercised and may be exercised in other cases'.[30] Brownlie concludes more generally that, whatever 'the particular and interstitial significance of equity in the law of nations, as a general reservoir of ideas and solutions for sophisticated problems it offers little but disappointment'.[31]

Equity as a general principle of law may play an important role in facilitating the rational and structured integration and reconciliation of different and competing objectives or values. Equity is, after all, commonly understood as a legal means of taking account of all relevant circumstances in a particular case, while the concept of sustainable development emerged in conventional and declaratory instruments of international environmental law as a means of reconciling protection of the natural environment with the requirements of economic and social development. In relation to the use of shared freshwater resources, it is generally accepted that the principle of equitable and reasonable utilisation 'operationalises' the concept of sustainable development, as sustainability is a goal which

---

[30] Brownlie, *supra*, n. 25, at 287.
[31] *Ibid.*, at 288.

could only be attained by reliance on equity.[32] The flexibility inherent in the application of equitable principles give them particular utility 'where there are competing interests which have not hardened into specific rights and duties ... in areas where the law is not highly developed. The nascent concept of intergenerational equity, and of equitable principles in environmental law, are examples'.[33] Thus, equitable principles can be understood as a legal means of facilitating the integration of diverse values and objectives, including environmental, social and economic values, into implementation of the multi-faceted principle of equitable and reasonable utilisation. This role involves a non-controversial application of equity *infra legem*, 'which constitutes a method of interpretation of the law in force and is one of its attributes'[34] and has been characterised as equity 'used to adapt the law to the facts of individual cases'.[35] This role for equity is likely to become ever more apparent, and ever more important, as international water law struggles, not only to take account of the myriad factors potentially relevant to equitable and reasonable utilisation of the kind indicated in Article 6(2) of the UN Watercourses Convention, but also to take on board the burgeoning normative implications of the increasingly pervasive requirements of ecosystems protection under international law or of the emerging human rights of access to adequate water and sanitation.

3.4.3.2 Procedural equity

The procedural rules of international water law play a particularly significant role in relation to the equitable and reasonable utilisation of shared waters, because they facilitate the orderly collection and communication of information according to agreed processes and methodologies, which is absolutely vital to any equitable consideration of the respective interests of States. It is notable that in the 2010 *Pulp Mills* case[36] the ICJ recognised the central role of procedural rules in the meaningful and effective implementation of the substantive rules of international water law and,

---

[32]   C.B. Bourne, 'The Primacy of the Principle of Equitable Utilization in the 1997 Watercourses Convention', (1997) 35 *Canadian Yearbook of International Law* 215–232, at 221–230.
[33]   Lowe, *supra*, n. 28, at 73.
[34]   *Frontier Dispute* case *(Burkino Faso v Mali)* (1986) *ICJ Reports* 554.
[35]   M. Akehurst, 'Equity and General Principles of Law', (1976) 25 *International and Comparative Law Quarterly* 801–825, at 801.
[36]   *Supra*, n. 9.

more particularly, that the process of environmental impact assessment plays a key role in ensuring that social, economic and environmental considerations relating to a planned or continuing use of an international watercourse are adequately understood and communicated, so that they may properly be taken into account as a factor within the balancing process that lies at the heart of equitable and reasonable utilisation. Therefore, international water law tends to stress the equitable participation of all riparian States. For example, Article 4 of the UN Watercourses Convention seeks to ensure the right to participate of any watercourse State that might be significantly affected by a proposed watercourse agreement between other co-riparians, while Article 5(2) sets out the ancillary principle of 'equitable and reasonable participation'. Characteristically, while Articles 11–19 set out detailed rules on notification, consultation and negotiation in respect of 'planned measures', Article 13(b) allows for extension of the period for reply to notification 'at the request of a notified State for which the evaluation of the planned measures poses special difficulty'. This provision typifies the imperative of equitably ensuring the meaningful and effective participation of all watercourse States in the process of international water law, which can perhaps be regarded as a practical application of the equitable maxim 'equality is equity'.

It is well understood that a normative framework requiring the equitable balancing of the legitimate interests of basin States must inevitably involve intense procedural inter-State engagement, which often can only be facilitated by the establishment of technically competent inter-State institutional machinery. Such institutions can ensure effective inter-State communication which might involve, inter alia, prior notification of planned projects potentially impacting upon the watercourse, routine exchange of information regarding the utilisation or condition of the shared waters, or expression of concerns on the part of any basin State. The pivotal role of institutional mechanisms in giving effect to the principle of equitable and reasonable utilisation has long been recognised by the international community, with Recommendation 51 of the *Action Plan for the Human Environment* adopted at the 1972 Stockholm Conference calling for the 'creation of river basin commissions or other appropriate machinery for cooperation between interested States for water resources common to more than one jurisdiction', and setting down a number of basic principles by which the establishment of such bodies should be guided.[37] Although such institutional structures can take numerous different forms and have diverse remits, there are today at least 119 river basin organisations (RBOs) performing a very extensive range of coordination and joint management functions.[38] Reliance on such institutional mechanisms to facilitate the inter-State cooperation necessary to achieve equitable and reasonable utilisation is often referred to as the 'common management' approach, which further underlines the existence of a community of interest among co-basin States. Consistent with the principle of equitable and reasonable participation articulated under Article 5(2) of the UN Watercourses Convention, the rules of international water law ought to be interpreted and applied in such a manner as to require that basin States take all reasonable measures to facilitate the meaningful and effective participation of other basin States in such cooperative institutional mechanisms. In some circumstances, this might require, for example, the provision of financial or technical assistance in order to ensure a less capacitated State's equitable participation.

3.4.3.3 Equity as a substantive rule of apportionment

When viewed as a stand-alone, substantive rule for the apportionment amongst States of the uses or benefits of shared international natural resources, three possible roles have been identified, which may be complementary and may occur concurrently: i.e., equity as a means of achieving a desired equitable result; equity as a process for taking account of all the relevant circumstances in a particular case; and equity as a means of making laws of general application more specific and readily applicable to a particular case.[39]

One of the roles of equity most widely employed by international courts and tribunals is that of choosing among possible interpretations of the law in such a way as to reach a just or equitable solution. The ICJ took this approach in the *Tunisia-Libya Continental Shelf* case,[40] where the Court

---

[37] UNCHE, *Report of the UN Conference on the Human Environment* (5–16 June 1972) UN Doc A/CONF.48/14/Rev.1, at 17.
[38] S. Schmeier, *Governing International Watercourses: River Basin Organisations and the Sustainable Governance of Internationally Shared Rivers and Lakes* (Routledge, Abingdon, 2013), at 65.
[39] R. Higgins, *Problems and Process: International Law and How We Use It* (Clarendon Press, Oxford, 1994), at 220–222.
[40] (1982) *1CJ Reports* 18.

focused on achieving what it regarded as an equitable result. However, the Court insisted that the search for an equitable result was not an operation of distributive justice but merely an operation of equity in a corrective role. This corrective function can only take place in a manner consistent with the rules of law and would never be acceptable *contra legem*. In the *Libya-Malta* case,[41] the ICJ again reiterated the distinction between this role of equity and the operation of distributive justice. However, while this argument might be tenable in maritime boundary delimitation, it is a great deal more difficult to argue that the act of interpreting the rules of international water law so as to achieve an equitable result was not an operation of distributive justice. The cardinal and overarching principle of equitable and reasonable utilisation has long been understood as a process for the allocation of rights in the uses and benefits of shared water resources on the basis of a distributive conception of equity having regard to all relevant factors, whereby uses and benefits will be shared in proportion to each basin State's needs and such needs are calculated through consideration of factors such as those listed under Article 6(2) of the UN Watercourses Convention.

Higgins' second conception of equity suggests that equity in international law lacks specific content but operates rather as a means for considering all the relevant circumstances in a particular case. In the *Tunisia-Libya Continental Shelf* case,[42] the ICJ seems to have supported this view, finding that it was 'virtually impossible to achieve an equitable solution to any delimitation without taking into account the particular relevant circumstances of the area'. In this context, it would appear that 'there are few, if any, constraints upon the factors which may form the basis of an argument in equity' as 'there is no legal limit to the considerations which States may take account of for the purpose of making sure they apply equitable procedures'.[43] Emphasising the potential flexibility of equity in this role, Lowe further observes that 'once the relevant factors have been considered the person making the decision is freed from the necessity of making the reasoning consistent with established legal rules and principles', though he does concede that 'even equity must be consistent'. Equity in this role resonates with equitable and reasonable utilisation which, as formulated under the Helsinki Rules and the UN Watercourses

---

[41] (1985) *ICJ Reports* 13.
[42] *Supra*, n. 40.
[43] Lowe, *supra*, n. 28, at 72–73.

Convention, provides a non-exhaustive, indicative checklist of factors which are to be considered. However, neither formulation offers any guidance as to the weight or priority to be given to the various factors listed as relevant, so that the principle gives little normative guidance as to what should happen in any particular situation. Instead, each provides, rather unhelpfully, that all the factors must be balanced with other factors and a decision made on the basis of the whole. Due to its normative vagueness, some have tended to be pessimistic about the principle's usefulness, though others feel it retains merit as a procedural approach.

The third role for equity of relevance to the application of the principle of equitable and reasonable utilisation is that of establishing the specific content of rules which are too general or vague to be applied directly in certain circumstances. In this way, equity permits the application of general legal rules to specific, concrete situations. Therefore, equity might be expected to play a crucial role in elaborating the substantive content of the principle of equitable and reasonable utilisation. Looking more particularly at the application of equitable principles to shared natural resources, Franck recognises three distinct approaches to equitable allocation of shared resources: i.e., 'corrective equity'; 'broadly conceived equity'; and 'common heritage equity'.[44] Under the 'corrective equity' approach, equitable considerations are only exceptionally invoked and function to ameliorate the gross unfairness which might occasionally result from the strict application of technical legal rules. This is the most conservative approach, confining the exceptional application of equitable principles within a dominant rule of resource allocation. Under the 'broadly conceived equity' approach, equity itself comprises a rule of law and is the dominant applicable rule for resource allocation. This approach affords tribunals a great deal more discretion than corrective equity and tends to be more openly distributive. Franck regards the principle of equitable and reasonable utilisation as a prime example of broadly conceived equity and as indicative of a recent trend to include similar equitable mechanisms in natural resource and environmental treaty regimes. Generally, Franck identifies a trend (exemplified by the adoption of Article 83(1) of the 1982 UN Convention on the Law of the Sea, requiring states 'to achieve an equitable solution' in continental shelf delimitation) towards the introduction of broadly conceived equity into conventional

---

[44] T. Franck, *Fairness in International Law and Institutions* (Clarendon Press, Oxford, 1995), at 57.

provisions relating to allocation of shared natural resources, which will increasingly compel courts and tribunals to apply broader notions of distributive justice. 'Common heritage equity' applies to the allocation of resources which are the patrimony of all humanity such as outer space, Antarctica or the mineral resources of the deep seabed, and often involves a 'trust' model in which conservation is the first or sole priority. Clearly this last approach has limited relevance for shared freshwater resource, where utilisation rights are associated with territorial sovereignty over a portion of the watercourse or basin.

From the above examination of the potential roles of equity in the application of equitable and reasonable utilisation it is apparent that equity can, and often will, play a combination of the roles identified. Thus, it appears axiomatic that equity may be applied at several stages, i.e., in the identification of a just and equitable solution, in the consideration of all relevant factors and circumstances, and in the concrete elaboration of normatively vague rules. However, no formulation of the principle offers guidance as to the order in which equity might perform these functions. For instance, would a court identify a just result and go on to achieve this result by giving appropriately weighted consideration to each relevant factor, or would a just result be determined by prior consideration of the relevant factors? The former approach would afford a court a greater degree of discretion in establishing what constitutes an equitable result, which would necessarily be subjective. Similarly, where equity functions to elaborate specific rules for a particular case, it is unclear whether these rules determine the priority to be given to each relevant factor or whether prior consideration of the factors is necessary in order to determine the rules to be applied. What is clear is that equitable and reasonable utilisation can only effectively function as a procedure, a fact long recognised by expert commentators.

3.4.3.4 *Equity as proportionality*

In the process of territorial delimitation and shared resource allocation, the concept of proportionality has long been understood as 'one technique among many to achieve an equitable outcome in the face of special geographic circumstances'.[45] This is particularly the case in the history of continental shelf and maritime delimitation, where coastal States in

---

[45] Higgins, *supra*, n. 39, at 230.

dispute were called upon to recognise a reasonable degree of proportionality as between 'the extent of the continental shelf appertaining to the States concerned and the lengths of their respective coastlines'.[46] Rejecting the use of equity to promote distributive justice, the ICJ held that the role of equity, and of proportionality as an element of equity, was to correct geographical anomalies, in this case the concavity of (then) West Germany's coastline, rather than to ensure fair or equal shares.

However, whilst the approach taken by international tribunals to the role of proportionality in maritime delimitation is quite narrow and limited, the concept plays a more fundamental and far-reaching role with regard to shared water resources. International water law differs from that on continental shelf delimitation in that it has been thoroughly codified, thereby giving a solid legal basis to those considerations against which proportionality, as a function of equity, can be measured. It is significant that similar formulations of the principle of equitable and reasonable utilisation, complete with indicative lists of relevant factors, have received widespread support in State practice. One might assume from judicial practice in respect of continental shelf and maritime territory delimitation that, in identifying an equitable regime for the utilisation of an international watercourse, tribunals would emphasise the importance of the natural and physical characteristics of the watercourse within each State. Principal among such factors would be the length of the watercourse or extent of the drainage area lying within the territory of each State and/or the quantum contribution of water by each State to the flow of the watercourse. Indeed, the indicative list of factors relevant to equitable and reasonable utilisation provided under Article 6(1) of the UN Watercourses Convention suggests as much by listing as the first set of factors 'geographic, hydrographic, hydrological, climatic, ecological and other factors of a natural character'. However, a very comprehensive survey of State and judicial practice relating to the allocation of shared freshwaters concludes persuasively that the significance attributed to the geophysical characteristics of the drainage basin, and in particular the extent of the drainage area lying within the territories of the parties

---

[46] *North Sea* cases, *supra*, n. 20. See L.F.E. Goldie, 'Equity and the International Management of Transboundary Resources', in A. Utton and L. Teclaff (eds.), *Transboundary Resources Law* (Westview Press, London, 1987), at 118–119.

and their contribution of water to the flow of the river, is relatively low.[47] Fuentes points out, for example, that the Narmada Tribunal made a *prima facie* equitable apportionment on the basis of the social and economic needs of the parties, which was then modified slightly after consideration of the physical characteristics of the drainage basin, an accommodation which 'shows that the role ascribed to that [geophysical] criterion is low in the hierarchy of relevant factors'. She goes on to assert that this practice of attributing low significance to the physical characteristics of a drainage basin as a factor in establishing an equitable regime for the utilisation of shared water resources 'is consistent with the rule of equitable and reasonable utilisation, because to admit these factors to function as a direct basis for the allocation of waters would not be in keeping with the principle of equality between the basin States'. At any rate, by attaching more significance to the water needs of watercourse States than to the geophysical characteristics of the drainage basin, equity as applied in international freshwater law is considerably more distributive in nature than equity as applied in the law of continental shelf and maritime territorial delimitation.

3.5 Benefit-sharing

In elaborating upon the core obligation of watercourse States to 'utilise an international watercourse in an equitable and reasonable manner', Article 5(1) of the UN Watercourses Convention expressly stipulates that the watercourse 'shall be used and developed by watercourse States with a view to attaining optimal and sustainable utilisation thereof and benefits therefrom'. This emphasis on water-related benefits and on sustainable management of the watercourse suggests the role that so-called benefit-sharing arrangements might have to play in crafting a utilisation regime that is mutually equitable and reasonable, something that might otherwise prove elusive. Furthermore, as the advent of the ecosystem approach to the management of shared international watercourses creates greater awareness of the fact that downstream States benefit from ecosystem services safeguarded by upstream States, and that harm can flow in both directions in a watercourse with restrictions on water utilisation

---

[47] Fuentes, *supra*, n. 29, at 394–408.

impacting on upstream interests, benefit-sharing can help to reconcile the diverse interests of upstream and downstream States.

As the concept of benefit-sharing is not expressly recognised under any formal legal instrument, there exists no authoritative definition. However, the practice might be loosely defined to include any arrangement designed to change the allocation of costs and benefits associated with transboundary water cooperation, which would normally involve some form of redistribution or compensation. Benefit-sharing arrangements would typically involve some form of payments for benefits, or compensation for costs, associated with enhanced stewardship of a shared transboundary watercourse, normally undertaken by an upstream State and resulting in benefits for a downstream State(s). For example, a downstream State might make payments to upstream riparian States for improved watershed management practices designed to bring benefits downstream, such as improved water quality, managed environmental flows or reduced flooding. Such an arrangement regarding the stewardship of headwaters and watersheds would entitle upstream riparians to share some portion of the downstream benefits that their stewardship helps to facilitate, and thus require downstream riparians to share the costs of such stewardship. Benefits in this context might include a broad range of economic, social, environmental, and/or political gains. Thus, in situations where the simple allocation of a quantum share of water would prove inefficient, benefit-sharing arrangements would permit riparian States to cooperate in taking a basin-wide approach to the optimisation of water-related benefits and the allocation of associated costs, by providing a framework for the equitable sharing of those benefits and costs. The availability of such arrangements can function to encourage inter-State cooperation, whilst also greatly enhancing the range and scope of cooperative initiatives in which states might engage, by facilitating broad issue-linkage. In addition to benefits directly connected to water resources utilisation, such as irrigated food production or hydropower generation, benefit-sharing arrangements might also take account of non-water use-related benefits, such as increased trade or improved diplomatic relations.

The practice of benefit-sharing with respect to international water resources is generally understood to have its origins in the 1961 Columbia

River Treaty[48] concluded between Canada and the United States. This treaty provided for the construction and operation of three infrastructure projects in Canada designed to maximise benefits associated with hydropower generation capacity, irrigation and flood control in the United States. The treaty also dealt with compensation payments to Canada related to those benefits enjoyed by the United States. The International Joint Commission, established under the overarching 1909 Boundary Waters Treaty between Canada and the US, facilitated these discussions, which lasted many years. In order to inform negotiations, the two riparian States requested the Commission to develop a set of principles intended to govern any sharing of benefits. The agreed principles included the central requirements that any agreement 'should result in both the equitable sharing of the downstream benefits attributable to any cooperative undertakings that might take place, and an advantage to each country as compared to any alternatives that might be available to them'. Thus, it appears that benefit-sharing arrangements should essentially require that the project(s) or other forms of cooperation contemplated would benefit both or all riparian states insofar as such cooperation somehow 'enlarges the pie'.

The Columbia River example also suggests that, due to the novelty of benefit-sharing as a cooperative paradigm and the inherent complexity of the considerations and calculations involved, a sophisticated legal and institutional framework for cooperation would be required for formulating related proposals. In facilitating discussions over the detailed arrangements for the Columbia basin, the International Joint Commission established the International Columbia River Engineering Board, which in turn set up an Engineering Committee charged with 'obtaining data and analysing the situation'. This committee continued to carry out extensive technical studies for many years. The relative success of the Columbia River benefit-sharing regime can be contrasted with the problems experienced in attempts to introduce benefit-sharing in the Amu Darya and Syr Darya basins in Central Asia, where a binding legal framework and competent institutional arrangements were lacking. Of course, such elaborate institutional structures are commonly associated with the general obligation of watercourse States to cooperate in the

---

[48] 1961 Treaty Relating to Cooperative Development of the Water Resources of the Columbia River Basin (17 January 1961) 542 UNTS 244.

utilisation and protection of an international watercourse,[49] one of the generally applicable rules of international water law providing a legal basis for the practice of benefit-sharing.

3.5.1 Legal Basis

The practice of entering into formal inter-State arrangements for the broader sharing of benefits derived from the utilisation of the water resources of an international watercourse, rather than the narrower and more simplistic practice of volumetric allocation of the available transboundary waters, has variously been described as 'a shared benefits strategy' and as 'a principle of equitable sharing of downstream benefits', while it has also been argued that this practice 'has become a general principle of both international water law and environmental law'. While any lawful practice entered into by consenting sovereign States does not require legitimation by reference to an established rule or principle of general international law, it is still useful to investigate if such a practice is consistent with the key rules of international water resources law. Examination of the practice's relationship with such rules helps us to understand better the objectives, considerations and legal parameters that ought to guide States in crafting benefit-sharing arrangements. In this regard, it appears that benefit-sharing can rely for support upon all three of the key rules of general international water resources law, i.e., the principle of equitable and reasonable utilisation, the duty to prevent significant transboundary harm, and the general duty to cooperate in the utilisation and protection of an international watercourse.

Benefit-sharing provides an important means of giving practical effect to the universally accepted principle of equitable and reasonable utilisation, and the wording of Article 5(1) of the UN Watercourses Convention, widely regarded as indicative of customary international law, suggests the possibility that watercourse States may either allocate transboundary water resources volumetrically and/or agree to the sharing of benefits arising from shared watercourses. In other words, benefit-sharing arrangements might serve to substitute for or complement arrangements for allocating a quantum share of water resources. When employed by co-riparian States, benefit-sharing would typically require certain States to forgo a potentially beneficial use of the shared waters *in lieu* of mon-

---

[49] UN Watercourses Convention, art 8.

etary or other compensation for allowing other States to put the water to a more efficient use. Conceiving benefit-sharing as a means of giving effect to the principle of equitable and reasonable utilisation highlights the practice's general objective of achieving equitable apportionment of whatever additional benefits are to be achieved by means of enhanced inter-State cooperation. While what constitutes an 'equitable' share is necessarily vague, flexible and context-dependent, and is clearly even more so in the case of a range of water-related and/or non-water-related benefits than in the more straightforward case of water quantum, we nevertheless have some grasp of the function and application of equity in this context. Having regard to humans' unique dependence upon water, we increasingly recognise the significance of the equitable principle of proportionality in ensuring that equity as applied in international water law is more distributive in nature than equity as applied to other contested natural resources. The flexible and open-textured nature of the principle of equitable and reasonable utilisation is confirmed by the non-exhaustive list of factors relevant to its determination, which normally accompanies any detailed articulation of the principle. Though the list set out in Article 6 of the UN Watercourses Convention includes no express reference to benefit-sharing, cost-sharing or the payment of compensation, this list is quite clearly intended to be open-ended, and nothing precludes such arrangements being taken into account in reaching or identifying an equitable accommodation of the interests of co-basin States. In support of this contention, the list included in the seminally important 1966 Helsinki Rules appears to contemplate just such arrangements by expressly including, as a factor relevant to the determination of a reasonable and equitable share in the beneficial uses of the waters of an international drainage basin, 'the practicality of compensation to one or more of the co-basin States as a means of adjusting conflicts among uses'.[50]

Benefit-sharing arrangements would also tend to function as a means of giving effect to the second principal substantive rule of general international water law, the duty to 'take *all appropriate measures* to prevent the causing of significant harm to other watercourse States'.[51] Article 7(2) elaborates further on the nature of this obligation, requiring States to consult with affected States in eliminating and mitigating such harm and, where necessary, regarding payment of compensation. This authoritative

---

[50] *Supra*, n. 4, art V(II)(10).
[51] UN Watercourses Convention, art 7(1), (emphasis added).

formulation of the no-harm rule as applied to international watercourses therefore envisages the possibility of a State agreeing to activities within the territory of another State that might adversely impact upon the former's interests, along with some arrangement for compensation in respect of such adverse impact.

Environmental harm, and damage to the ecosystem of an international watercourse in particular, are among the key adverse impacts envisaged under Article 7 of the UN Watercourses Convention and under customary international law.[52] Article 20 expressly requires that '[w]atercourse States shall, individually and, where appropriate, *jointly*, protect and preserve the ecosystems of international watercourses', reflecting broad State practice and, almost certainly, the position in customary international law.[53] The normative nature and implications of ecosystem obligations are becoming increasing clear, based on sophisticated analytical and methodological concepts, such as those regarding environmental flow requirements and the evaluation of ecosystem services. The requirement for States to act jointly, where appropriate, in environmental protection could be taken to support the use of benefit-sharing arrangements under which downstream States agree to compensate upstream States in lieu of forgone water resources development opportunities or for the adoption of measures aimed at protecting and preserving international watercourse ecosystems to the benefit of such downstream States. Of course, harm to the interests of watercourse States does not only 'travel downstream', as prior water utilisation by downstream States, might result in the foreclosure of upstream States' intended future water use. One examination of benefit-sharing arrangements acknowledges that 'downstream extraction generates externalities upstream by diminishing future flows available for abstraction upstream, by virtue of perceptions of acquired rights to that water downstream'.[54] Thus, such 'foreclosure' of water utilisation

---

[52] See also, International Law Association, *Report of the Seventy-First Conference of the International Law Association* (2004), art 8 ('Berlin Rules'); UNECE Water Convention, *supra*, n. 8, art 2.

[53] Watercourse States are also required, where appropriate, to act 'jointly' under art 21 (Prevention, Reduction and Control of Pollution) and art 23 (Protection and Preservation of the Marine Environment) of the UN Watercourses Convention.

[54] C.W. Sadoff and D. Grey, 'Cooperation on International Rivers: A Continuum for Securing and Sharing Benefits', (2005) 30(4) *Water International* 420–442, at 424. See also, S.M.A. Salman, 'Downstream

opportunities for upstream States might amount to harm for the purposes of the no-harm rule, meaning that benefit-sharing arrangements involving some form of compensation for upstream States which forgo such opportunities for the sake of ecosystems protection, or to facilitate more efficient downstream use, would be entirely consistent with the no-harm principle, especially in the light of the express reference to the possibility of compensation in Article 7(2) of the UN Watercourses Convention.

Benefit-sharing arrangements also find legal support in the third key rule of general international water law, the general duty of watercourse States to cooperate in the utilisation and environmental protection of an international watercourse. This obligation is largely comprised of procedural elements, including the well-established duties to notify co-riparian states of planned projects, to consult and, where necessary, negotiate with co-riparian States where disagreements arise, and to engage in the regular exchange of relevant water-related information. The duty to cooperate is generally regarded as essential in facilitating meaningful implementation of the two key substantive obligations, i.e., equitable and reasonable utilisation and the prevention of significant transboundary harm. Article 8 of the UN Watercourses Convention, on the 'general obligation to cooperate', can be considered an authoritative indication of that norm's position in customary international law. It establishes that 'watercourse States shall cooperate on the basis of sovereign equality, territorial integrity, *mutual benefit* and good faith in order to attain *optimal utilization* and *adequate protection* of an international watercourse' (emphasis added). This emphasis suggests that benefit-sharing arrangements under which upstream States are rewarded for maintaining the ecosystem of an international watercourse fall within the scope of the cooperative initiatives envisaged. In elaborating upon the precise legal origins and meaning of Article 8, the International Law Commission commentary to its 1994 Draft Articles explains that inter-State watercourse cooperation 'is an important basis for the attainment and maintenance of an equitable allocation of the uses *and benefits* of the watercourse'.[55] In providing an example of an instrument incorporating typical provisions on transboundary water coopera-

---

Riparians Can Also Harm Upstream Riparians: The Concept of Foreclosure of Future Uses', (2010) 35(4) *Water International* 350–364.
[55]   *Supra*, n. 2, at 105 (emphasis added).

tion,[56] the Commission corroborates the understanding that cooperation under Article 8 might include joint arrangements incorporating elements of benefit-sharing. Consistent with the experience of the Columbia River basin States in crafting a benefit-sharing regime, Article 8(2) of the UN Watercourses Convention suggests that 'watercourse states may consider the establishment of joint mechanisms or commissions, as deemed necessary by them, to facilitate cooperation on relevant measures and procedures'. The key function of any joint watercourse mechanism or commission is that of facilitating the regular and structured exchange of data and information, acknowledged as an absolutely critical prerequisite for the conclusion of any benefit-sharing arrangements. Meaningful and sophisticated inter-State information exchange is necessary 'to reach common understanding, change perceptions, and achieve information symmetry, in order to build trust and catalyse cooperation'.[57] More generally, a range of universally adopted declarative instruments can be understood as an acknowledgement of the need for more sophisticated cooperative arrangements within the framework of international law, in order to ensure that environmental costs are fully and equitably internalised, including by means of benefit-sharing and eco-compensation arrangements.[58] Of course, it must also be borne in mind that the authoritative articulation of the principle of equitable and reasonable utilisation contained in the UN Watercourses Convention, includes the direction to watercourse States to 'participate in the use, development and protection of an international watercourse in an equitable and reasonable manner',[59] thereby emphasising the critical role of inter-State cooperation in 'attaining optimal and sustainable utilization thereof and benefits therefrom',[60] by whatever means might be appropriate.

---

[56] 1964 Agreement between Poland and the USSR Concerning the Use of Water Resources in Frontier Waters (17 July 1964) 552 UNTS 175, arts 3(3), 7 and 8.
[57] Sadoff and Grey, *supra*, n. 54, at 425.
[58] UNCHE, Stockholm Declaration, *supra*, n. 37, Principle 22; UNCED, Rio Declaration, *supra*, n. 13, Principles 13 and 24.
[59] UN Watercourses Convention, art 5(2).
[60] *Ibid.*, art 5(1).

3.6 Solidarity and equity in international water law

The concept of 'solidarity', long mooted as 'a fundamentally sound principle of international law',[61] and understood by some 'to be the basic condition for the existence of a community of states',[62] provides a useful lens through which to consider the role of equity in international water law. As a principle founded upon recognition of the essential interdependence of the community of States, and thus the need for meaningful cooperation amongst States in order to maximise welfare, it has clear relevance for the progressive development of international environmental and natural resources law, where solidarity can act as a possible counterweight to narrow and self-interested conceptions of States' rights flowing from territorial sovereignty.[63] The historical evolution of international water law has been largely based upon contested conceptions of territorial sovereignty and jealous protection of the uses and benefits based thereon, which has hindered cooperative inter-State efforts to address the natural tensions arising between upstream and downstream watercourse States and between more and less developed watercourse States. Sovereignty, therefore, regularly operates to retard meaningful inter-State transboundary water cooperation.[64] While such opposing positions reflect the immediate, short-term self-interest of States in contrasting circumstances, the increasingly urgent imperative of cooperative management of shared international freshwater resources, with a view to optimising the welfare benefits derived from an increasingly scarce and environmentally vulnerable resource, presents a classic example of a collective action problem,[65] requiring co-basin States to engage in intensive cooperative efforts

---

[61] R.St.J. Macdonald, 'Solidarity in the Practice and Discourse of Public International Law', (1996) 8(2) *Pace International Law Review* 259–302, at 259; R. Wolfrum, 'Solidarity and Community Interests: Driving Forces for the Interpretation and Development of International Law', (2021) 416 *Recueil des Cours* 9–479.
[62] E. De Vattel, *Le Droit des Gens ou Principes de la Souveraineté* (1758, reprinted Sweet, Stevens & Maxwell, London, 1958).
[63] S.C. McCaffrey, *The Law of International Watercourses* (2nd ed.) (Oxford University Press, Oxford, 2007), at 67–68.
[64] J. Brunnée and S.J. Toope, 'Environmental Security and Freshwater Resources: A Case for International Ecosystem Law', (1994) 5 *Yearbook of International Environmental Law* 41–76, at 56.
[65] E. Benvenisti, 'Collective Action in the Utilization of Shared Freshwater: The Challenges of International Water Resources Law', (1996) 90 *American Journal of International Law* 384–415.

involving difficult compromise on the part of all States. Such needs reflect general international law's continuing functional evolution from a 'law of co-existence' to a 'law of cooperation'.[66] Therefore, the concept of solidarity, which is intrinsic to transboundary water governance offers the prospect of an organising principle which might assist in assuaging the more confrontational implications of the principle of sovereignty by engendering mutual understanding amongst riparian States with differing needs and water security concerns, so as to ensure 'reconciliation of conflicts of interest with a solidarity-based balancing of human livelihood interests ... to be achieved against unavoidable environmental consequences'.[67] In the particular context of shared international water resources, solidarity may assist formal recognition of the common interests and interdependence of co-basin States as a foundation for collective action which has regard to the concept of sustainable development and employs a holistic approach to the basin, with the ultimate aim of achieving optimal resource use, more equitable distribution of water-related benefits and sustainable outcomes. It is best understood as a counterweight to more traditional paradigms of international water resources management, and in particular to more uncompromising conceptions of territorial sovereignty. At its core, so called 'hydro-solidarity' seeks to introduce an ethical element to transboundary water management, stressing interdependency and the need for compromise in order to optimise human livelihood interests and effective ecological protection. In pursuit of such aims, regarded as central elements of the guiding concept of sustainable development, hydro-solidarity requires robust institutions—both in the form of rules and organisations—to facilitate effective exchange of information and broad stakeholder participation in order to identify common goals and make difficult compromises socially acceptable.[68] Therefore, solidarity can operate to restrain unilateral action and to promote inter-State compromise based on sophisticated benefit-sharing arrangements which respect the hydro-climatic and socio-ecological limits of the shared basin. Commentators identify a number of elements that can assist in facili-

---

[66] W. Friedman, *The Changing Structure of International Law* (Columbia University Press, New York, 1964).

[67] M. Falkenmark and C. Folke, 'The Ethics of Socio-Ecohydrological Catchment Management: Towards Hydrosolidarity', (2002) 6 *Hydrology and Earth System Sciences* 1–10, at 4.

[68] J. Gjørtz Howden, *The Community of Interest Approach in International Water Law: A Legal Framework for the Common Management of International Watercourses* (Brill Nijhoff, Leiden, 2020), at 58–59.

tating consideration of hydro-solidarity, including the use of extensive cross-sectoral or 'broad' information in decision-making, cooperative institutional machinery specifically designed to assist in finding compromise, the use of broad public participation in order adequately to address the social implications of resource use, and means for redressing resource use that damages the interests of other users.[69]

Despite few express endorsements in formal legal instruments, the broader concept of solidarity penetrates the closely inter-related fields of international water, environmental and natural resources law by virtue of the overarching objective of sustainable development, within which the concept of equity plays an absolutely central role.[70] One key dimension of equity that permeates the vision of sustainable development set out in the 1992 Rio Declaration is that of intra-generational equity, in the particular form of 'common but differentiated responsibilities' (CBDR) as between States bearing different degrees of responsibility for environmental degradation and possessing different national capacity to contribute to solutions.[71] Such differentiation represents a clear practical application of the concept of solidarity, which informs every aspect of the former. As one leading author explains regarding differentiation in the context of international environmental law, 'the measure of common responsibility in the CBDR principle is thus rooted in the principle of cooperation, which posits that states are obliged, in the spirit of solidarity, to cooperate in preventing transboundary pollution'.[72] CBDR is now widely accepted and increasingly well understood as a means of ensuring universal participation by States in initiatives aimed at achieving sustainable development by ensuring flexibility having regard to three key factors—need, capability and responsibility. Therefore, differentiated approaches influence key aspects of international water law practice in order to ensure that the concept of solidarity may inform the determination of interests in accordance with the distributive equity which permeates this sub-field

---

[69] B.O. Magsig, *International Water Law and the Quest for Common Security*, (Routledge, London, 2015), at 69.

[70] P. Wouters, 'The Relevance and Role of Water Law in the Sustainable Development of Freshwater: From "Hydrosovereignty" to "Hydrosolidarity"', (2000) 25 *Water International* 202–207.

[71] UNCED, Rio Declaration, *supra*, n. 13, Principle 7. See also, Principles 3, 4, 5 and 6 regarding the right to development, poverty alleviation, etc.

[72] L. Rajamani, *Differential Treatment in International Environmental Law* (Oxford University Press, Oxford, 2006), at 134.

of international law.[73] In line with the central tenets of the concept of solidarity, a distributive conception of equity involves the distribution of benefits derived from scarce shared resources primarily on the basis of the essential needs of each party.

The International Law Commission's 2008 Draft Articles on Transboundary Aquifers offer an unusual example of the express articulation of core CBDR values in Article 16 covering 'technical cooperation with developing States', which exhorts States, either acting alone or through an international organisation, to promote scientific, educational, legal and other cooperation with developing States for protection and management of transboundary aquifers.[74] Article 16 sets out in detail which forms such cooperative assistance to developing States should take, including capacity-building and enhancement, facilitating international participation, supplying equipment and facilities, providing technical and practical advice, and supporting exchange of technical knowledge and experience. Though the concept of hydro-solidarity does not enjoy, at least as yet a great deal of express support in formal international law instruments and might, therefore, be regarded by sceptics as a utopian paradigm, a closer examination of key normative elements of international water law and related State practice through the prism of solidarity suggests the role of values inherent to solidarity within this rapidly developing corpus of rules and principles. Most notably, the principle of equitable and reasonable utilisation is imbued with a profoundly distributive conception of equity recognising the need to identify and cater to the true water-related needs and vulnerabilities of each watercourse State. Among the factors commonly listed as relevant to any determination of the entitlements of each watercourse State are included 'the social and economic needs of the watercourse States concerned' and 'existing and potential uses of the watercourse',[75] with neither of the latter classes of use

---

[73] See O. McIntyre, 'Embedding "Solidarity" in International Water Law: Framing "Equity" in Transboundary Water Governance', (2020) 51 *Netherlands Yearbook of International Law* 227–256.

[74] International Law Commission, *Report of the International Law Commission on the Work of its Sixtieth Session*, UN GAOR, Sixty-Second Sess., Supp. No 10, (2008) UN Doc A/63/10.

[75] United Nations Convention on the Non-Navigational Uses of International Watercourses, (1997) 36 *International Legal Materials* 700 ('UN Watercourses Convention'), art 6.

given any inherent priority.[76] Such implicit recognition of varying degrees of social and economic dependence and vulnerability, and of the imperative of belated development, appears to echo the values inherent to the CBDR principle. Indeed, even though Article 6 of the UN Watercourses Convention lists 'economy of use of the water resources of the watercourse' as a factor relevant to equitable and reasonable utilisation, it also includes 'the costs of measures taken to that effect', thus having regard to watercourse States' capacity to bear such costs. Further, exhaustive examination of the relative weight attributed in State and judicial practice to each of the factors relevant to equitable and reasonable utilisation suggests that the importance of efficient utilisation is, in any case, quite limited.[77] It is telling, as regards the integration of solidarity values, that the only factor expressly acknowledged in the leading international water law instruments as enjoying priority in the determination of equitable and reasonable use is that of 'the requirements of vital human needs'. This all suggests that the inclusive and distributional instincts upon which CBDR is based have long guided the development of international water law.

Acknowledging the structural tensions at the heart of the traditional framework of international water law, international tribunals have long recognised the 'community of interest' existing among co-basin States,[78] a notion which appears to incorporate many of the values inherent to the concept of solidarity. A Permanent Court of Arbitration tribunal recognised this inter-linkage in the 2004 *Rhine Chlorides Arbitration*, where it acknowledged the community of interest existing among the Rhine States in addressing chlorides pollution, in which solidarity is an important factor.[79] The use of various kinds of benefit-sharing arrangements in transboundary basins also appears to imply the importance of considerations of solidarity, including the CBDR principle, in achieving equitable, optimal and sustainable utilisation of international water resources. As

---

[76] Commentary to the ILC Draft Articles, *supra*, n. 2, at 233.
[77] See Fuentes, *supra*, n. 29.
[78] *Case Relating to the Territorial Jurisdiction of the International Commission of the River Oder* (Judgment No 16) (1929) *PCIJ Reports*, Series A No 23, para 74; *Gabčíkovo-Nagymaros* case, *supra*, n. 1, para 85; *Pulp Mills* case, *supra*, n. 9, para 281.
[79] *Case Concerning the Auditing of Accounts between the Kingdom of Netherlands and the French Republic pursuant to the Additional Protocol of 25 September 1991 to the Convention on the Protection of the Rhine against Pollution by Chlorides of 3 December 1976* (2004) PCA, para 97.

the 'common interests' which CBDR serves to pursue may be increasingly concerned with solidarity between differently capacitated States and the goal of universal realisation of human rights,[80] the UN Sustainable Development Goals (SDGs) are increasingly likely to inform the elaboration, interpretation and application of norms of international water law and thereby emphasise differentiated approaches. Consider, for example, the sophisticated, yet contextually determined, targets and indicators identified under SDG 6, on access to water and sanitation for all, or under SDG 8, on inclusive and sustained economic growth.

Though CBDR as articulated in Principle 7 of the Rio Declaration suggests that it is only relevant to 'global environmental degradation', where 'developed countries acknowledge the responsibility that they bear', the principle's origins in promoting inclusive cooperative participation in a spirit of solidarity would appear to suggest that it is also relevant at other spatial scales, including at the level of transboundary river basins. This has long been implicit in the application of key rules of international water law. For example, the long-accepted variance in standards of due diligence expected of States (under the substantive rules of international water law, as elsewhere), having due regard to their relative stage of economic and social development and technical and administrative capacity, provides a clear and practical illustration of the values underlying the CBDR principle.

It is quite clear, therefore, that the concept of solidarity, and the related CBDR principle, influence operation of the rules of international water law, and play an important role in facilitating effective cooperation amongst transboundary watercourse States in addressing the looming global water crisis. As water scarcity, climate-related challenges and pollution problems come to impact ever more significantly on the management of international watercourses, inclusive participation by all watercourse States in increasingly ambitious cooperative arrangements will be required in order to identify and deliver truly equitable and sustainable water utilisation (and protection) regimes. In this regard, solidarity and differentiation offer to enhance significantly the conceptual vocabulary available to give expression to the common language of equity in international water law.

---

[80] B. Simma, 'From Bilateralism to Community Interest in International Law', (1994) 250 *Recueil des Cours* 218–384, at 233–243.

3.7 Conclusion

Thus, while the cardinal, overarching substantive principle of international water law is necessarily somewhat normatively vague and indeterminate, it essentially requires that each watercourse State, in its use of the resources of the watercourse, has due and adequate regard to the needs of other watercourse States and, in particular, to the nature and extent of their dependence upon the shared waters in question. Of course, this immediately suggests the critical importance of mechanisms for inter-State communication and cooperation, in order that States might adequately understand and take account of such dependence. As the looming global water crisis unfolds, it becomes ever more apparent that equitable and reasonable utilisation involves rather more than the equitable sharing of water quantum but envisages sophisticated benefit-sharing arrangements which can serve to optimise the benefits to be derived from a finite resource while safeguarding watercourse ecosystems and the essential ecosystem services provided thereby. In addition, the concept of solidarity is likely to play an ever-greater role in informing and framing the notion of equity in the face of the impending water, climate and biodiversity crises and persistent inequality in the capacity of watercourse States to respond and adapt. Such arrangements will require ever more sophisticated and robust inter-State cooperative mechanisms, implemented by properly mandated, technically competent and adequately resourced institutions.

4. Prevention of significant transboundary harm

4.1 Introduction

Whereas equitable and reasonable utilisation provides the cardinal, overarching rule of international water law, almost all international water resources agreements and codifications include a closely related obligation on watercourse States not to cause significant harm to other watercourse States. The existence of such a rule in general international law is supported by a wealth of authority in State practice, and formulations are included in both the 1997 UN Watercourses Convention[1] and the 1992 UNECE Water Convention.[2] This rule was recognised as established customary international law in 1941 by the arbitral tribunal in the Trail Smelter Arbitration and included among the general principles of nascent international environmental law by the 1972 Stockholm Declaration on the Human Environment.[3] In the specific context of international water law, very many watercourse agreements contain provisions on the prevention and abatement of water pollution and a representative formulation of the obligation of prevention was included in the seminally important 1966 Helsinki Rules.[4] A survey of watercourse agreements reveals a range of related ancillary substantive provisions dealing with, inter alia, the

---

[1] Convention on the Law of the Non-Navigational Uses of International Watercourses (adopted 21 May 1997, entered into force 17 August 2014), (1997) 36 *International Legal Materials* 700, art 7.
[2] 1992 UNECE Convention on the Protection of Transboundary Watercourses and International Lakes (adopted 17 March 1992, entered into force 6 October 1996) 1936 UNTS 269, art 2.
[3] UNCHE, *Report of the UN Conference on the Human Environment* (5–16 June 1972) UN Doc A/CONF.48/14/Rev.1, Principle 21.
[4] International Law Association, Helsinki Rules on the Uses of the Waters of International Rivers, ILA, *Report of the Fifty-Second Conference* (Helsinki 1966) 484, art X.

prevention of harmful effects, the protection of water quality, minimum flow requirements, and the application of clean technologies.

The requirement to prevent, however, is not absolute. While the obligation to prevent significant transboundary harm has obvious origins in territorial sovereignty and the doctrine of sovereign equality of States, it is also reflective of a number of legal maxims and doctrines upon which national legal frameworks are often based, including the Roman law maxim *sic utere tuo ut alienum non laedas* (so use your own [property] as not to harm that of another), the doctrine of abuse of rights (*abus de droit*; *Rechtsmissbrauch*), and the doctrine of good neighbourliness (*droit international de voisinage*; *Nachbarrecht*). Indeed, its close parallels with national rules firmly established in several of the world's principal legal systems might help to explain the obligation's universal acceptance by States. It is instructive that none of the above founding doctrines represents an absolute rule of prohibition. Instead, each attempts to reconcile conflicting rights in property or a related shared resource. Thus, the obligation to prevent significant transboundary harm can be understood as seeking to define the scope of States' rights in a way that is reasonable *vis-à-vis* other States, and thereby functions to moderate the effect of apparently absolute rules. Therefore, it is a due diligence obligation requiring the taking of reasonable measures by States in the use and protection of shared water resources, rather than an absolute prohibition on causing or permitting harm in all circumstances. As explained in the commentary to the International Law Commission's 1994 Draft Articles, which formed the basis for the 1997 UN Watercourses Convention, 'it is an obligation of conduct, not an obligation of result'.[5] Thus, a State might lawfully fail to prevent significant transboundary harm, provided that it had taken all reasonable measures to try to prevent such harm, which occurred despite that State's reasonable efforts. In the context of environmental conventions or treaty provisions, for example, the 'no-harm' rule doesn't tend to establish a strict obligation not to pollute (obligation of result), but only the obligation to endeavour to employ a due diligence of conduct in an effort to prevent, control and reduce pollution to the extent reasonably possible. Any breach of such an obligation will involve responsibility on the part of the State for fault, i.e., for lack of due diligence.

---

[5] International Law Commission, *Report of the International Law Commission on the Work of its Forty-Sixth Session*, UN GAOR, Forty-Ninth Sess., Supp. No 10, (1994) UN Doc A/49/10, at 237.

4.2 Normative content

The no-harm rule may be criticised as a rule that is often disregarded due to problems of uncertainty—both scientific uncertainty in relation to the causes of harm and legal uncertainty as to its interpretation and practical application. However, despite the principle's inherent flexibility and contextual relativity, it is possible to identify its key elements.

4.2.1 The Concept of Harm

As regards the concept of 'harm', it is clear that this might occur in a shared watercourse as a result of, inter alia, pollution, over-abstraction, obstruction of fish migration, erosion of riverbanks or riverbed, siltation, interference with the flow regime, or damage to riverine ecosystems. The concept of harm is understood broadly and the obligation to prevent harm is not confined to one State's direct use of shared watercourse causing harm to another State's use thereof. Global, regional and basin-level conventions tend to address a range of problems, such as maintenance of minimum flow requirements,[6] prevention of transboundary impacts[7] and the protection of water quality.[8] Harmful conduct may include activities in one State not directly related to a watercourse, such as deforestation, which may cause harmful effects in another State, such as excessive siltation or flooding. The International Law Association commentary to Article X of the 1966 Helsinki Rules notes that, for the purposes of the no-harm rule, 'an injury in the territory of a State need not be connected with that State's use of the waters'.[9] As regards the significance threshold for harm to be prevented under Article 7 of the UN Watercourses Convention, the International Law Commission has explained that '[t]here must be a real impairment of use, i.e., a detrimental impact of some consequence upon, for example, public health, industry, property, agriculture or the environment in the affected State'.[10] This

---

[6] 1995 Agreement on Cooperation for the Sustainable Development of the Mekong River Basin, (1995) 34 *International Legal Materials* 864, art 6.
[7] UNECE Water Convention, *supra*, n. 2, art 2.
[8] 1994 Treaty of Peace between Israel and Jordan, (1995) 34 *International Legal Materials* 43, art 3 Annex II.
[9] *Supra*, n. 4, at 500.
[10] International Law Commission, 'The Law of the Non-Navigational Uses of International Watercourses', (1988) 2(2) *Yearbook of the International Law Commission*, at 22–54.

approach formalises the so-called *de minimis* rule, which derives from the general principle of good neighbourliness and requires States to overlook small and insignificant inconveniences.

The continuing evolution of the so-called 'ecosystem approach' to the management of transboundary water resources, endorsed by Articles 20–23 of the UN Watercourses Convention, ought to ensure that Article 7 of the Convention is construed broadly, at least in relation to any ecological or environmental damage. Such an approach is evident in the application of rules of general international law by international courts and tribunals.[11] Today, impacts upon the functioning of aquatic ecosystems and/or loss of ecosystem services would certainly be included among the significant detrimental impacts upon the environment envisaged by the International Law Commission. It is quite clear that interference with the minimum environmental flow of an international watercourse could now be regarded as having caused significant harm.[12] Similarly, any material interference with or loss of ecosystem services provided by the riverine ecosystem of a shared watercourse may amount to actionable significant harm. In this regard the ICJ has found in respect of four categories of ecosystem services (out of twenty-two types of ecosystem services identified in the dispute, and six types for which compensation was claimed) that:

> These activities have significantly affected the ability of the two impacted sites to provide the above-mentioned environmental goods and services. It is therefore the view of the Court that impairment or loss of these four categories of environmental goods and services has occurred and is a direct consequence of Nicaragua's activities.[13]

4.2.2 Due Diligence Obligations

In describing the nature of the due diligence obligation of watercourse States confirmed under Article 7 of the UN Watercourses Convention, the

---

[11] For a notable example, see International Court of Justice, *Certain Activities Carried Out by Nicaragua in the Border Area (Costa Rica v Nicaragua)* and *Construction of a Road Along the San Juan River (Nicaragua v Costa Rica)* (Judgment on 16 December 2015) (2015) *ICJ Reports* 665.

[12] Permanent Court of Arbitration, *Indus Waters Kishenganga Arbitration (Pakistan v India)* [2013] (Partial Award, 18 February 2013), para 454; *Certain Activities* case, *ibid.*, paras 105 and 119.

[13] International Court of Justice (ICJ), *Certain Activities Carried Out by Nicaragua in the Border Area (Costa Rica v Nicaragua)*, (Compensation Judgment on 2 February 2018), para 75.

International Law Commission commentary to the 1994 Draft Articles refers approvingly to the definition of due diligence provided in the 1872 *Alabama Claims Arbitration*, which describes the concept as 'a diligence proportioned to the magnitude of the subject and to the dignity and strength of the power which is exercising it' and as 'such care as governments ordinarily employ in their domestic concerns'.[14] In the specific context of international watercourses, the 'magnitude of the subject' might be understood to refer to the nature of the relevant activity and suggests, therefore, that where it is inherently dangerous 'the care required would be so great as to approach strict liability; a virtual guarantee that such a harmful event would not occur'.[15] The appropriate measures to prevent the causing of significant harm required of each watercourse State under Article 7 of the UN Watercourses Convention can be understood to include both substantive and procedural elements, as the International Court of Justice made clear in the 2010 *Pulp Mills* case.[16] Substantive due diligence might require, for example, that a watercourse State with the potential to cause transboundary harm should ensure the adoption and enforcement of appropriate domestic regulatory controls on water use and pollution, while procedural due diligence might demand that a State planning a major project likely to impact on the watercourse or on the interests of co-basin States should notify such States and, where necessary, engage in good faith consultation and negotiation regarding their neighbouring States' outstanding concerns.

While it is quite clear that due diligence-based standards of conduct on the part of the State are absolutely central to any determination of the normative content of the no-harm rule, these same standards can prove to be rather abstract and elusive. 'Due diligence' is generally employed in international law to denote a notionally similar standard of care required in a range of diverse contexts.

According to one comprehensive study, the concept

> is concerned with supplying a standard of care against which fault can be assessed. It is a standard of reasonableness, of reasonable care, that seeks to

---

[14] *Supra*, n. 5, at 236–237.
[15] S.C. McCaffrey, *The Law of International Watercourses: Non-Navigational Uses* (Oxford University Press, Oxford, 2001), at 373.
[16] International Court of Justice, *Case Concerning Pulp Mills on the River Uruguay (Argentina v Uruguay)* (Judgment) [2010] *ICJ Reports* 14, para 77.

take account of the consequences of wrongful conduct and the extent to which such consequences could feasibly have been avoided.[17]

Aside from its inherent flexibility, the due diligence standard allows States a degree of autonomy, which coheres with ideas of sovereign discretion and might generally be expected to encourage wider participation in treaty regimes and constructive engagement with customary requirements. Of course, the open-ended nature of standards of due diligence also offers convenience, obviating the need to agree precise international rules, which might prove very difficult in practice, and may even prove premature where State practice and practicable standards are still evolving. One can view due diligence as 'a technique of proceduralisation, deferring controversial inquiries as to the content of substantive rules regulating wrongdoing to less controversial questions relating to informed decision-making and process'.[18] More generally in this regard, Koskenniemi notes the prevalence of 'contextual determinants ... in respect of rules of State responsibility and especially the customary standard of due diligence', the recent use of which he associates with 'the search for equitableness [which] has affected the law on, for example, natural resources'. This search for 'equitableness' might be regarded as the defining characteristic of international water law.

4.2.2.1 Generally applicable standards

In its 2001 Draft Articles on the Prevention of Transboundary Harm from Hazardous Activities, the International Law Commission identifies a general duty of prevention, comprising a due diligence obligation on the State of origin to take all reasonable preventive and/or mitigating measures.[19] Draft Article 3 provides that 'the State of origin shall take all appropriate measures to prevent significant transboundary harm or at any event to minimize the risk thereof' and reflects general State practice, particularly practice in the field of international environmental law. Referring expressly to Draft Article 7 of the Commission's 1994

---

[17] International Law Association Study Group on Due Diligence in International Law, *Second Report* (2016).
[18] M. Koskenniemi, *From Apology to Utopia: The Structure of International Legal Argument* (Cambridge University Press, Cambridge, 1989).
[19] International Law Commission (ILC), Articles on the Prevention of Transboundary Harm from Hazardous Activities, *Report of the International Law Commission*, Fifty-Third Sess. (2001), UN Doc A/56/10, at 148–170.

Draft Articles on the Law of the Non-Navigational Uses of International Watercourses, the commentary to Draft Article 3 provides a very clear account of the normative nature of this firmly established customary obligation:

> The obligation of the State of origin to take preventive or minimization measures is one of due diligence. It is the conduct of the State of origin that will determine whether the State has complied with its obligation under the present articles. The duty of due diligence involved, however, is not intended to guarantee that significant harm be totally prevented, if it is not possible to do so. In that eventuality, the State ... [must] exert its best possible efforts to minimize the risk. In this sense, it does not guarantee that the harm would not occur.

This understanding of the nature of the duty to prevent is entirely consistent with the position put forward in the UN Watercourses Convention, and generally considered to be reflective of customary international water law, whereby prevention is secondary to, and subordinated to equitable and reasonable utilisation, the overarching cardinal principle of international water law. Therefore, where a particular use of shared water resources represents the most equitable and reasonable allocation of the benefits deriving therefrom having regard to all relevant circumstances, any resulting harm to another watercourse State may have to be tolerated, though every effort should be made to minimise such harm and appropriate compensation might be due to the injured State.[20]

According to the International Law Association Study Group on Due Diligence, '"reasonableness" is a golden thread in determining which measures States should take to act in a duly diligent manner', and one commentator describes due diligence as 'a flexible reasonableness standard adaptable to particular facts and circumstances'.[21] It is clear that 'reasonableness' is equally central in determining a reasonable use for the purposes of the principle of equitable and reasonable utilisation.[22] Closely linked to the general standard of reasonableness, is the expectation of 'good government', which suggests that the due diligence standard

---

[20] UN Watercourses Convention, *supra*, n. 1, art 7(2).
[21] R.P. Barnidge, 'The Due Diligence Principle under International Law', (2006) 8(1) *International Community Law Review* 81–121.
[22] A. Rieu-Clarke, R. Moynihan and B.O. Magsig, *UN Watercourses Convention: User's Guide* (UNESCO IHP-HELP Centre for Water Law, Policy and Science, Dundee, 2012).

expected would involve 'the reasonable measures of prevention which a well-administered government could be expected to exercise under similar circumstances'.[23] Of course the reasonableness of any such expectation would be qualified to some degree by consideration of the State of origin's level of development. In turn, the linked notions of good government and level of development are connected to the degree of effective control which a State of origin is in a position to exercise over its territory and over non-State actors operating therein.

The commentary to the International Law Commission's Draft Article 3 on Prevention provides some broad guidance on the normative parameters of 'an obligation of due diligence as the standard basis for the protection of the environment from harm', advising, for example, that it requires policies which 'are expressed in legislation and administrative regulations and implemented through various enforcement mechanisms'. Consistent with the *Alabama* case, however, it makes it quite clear that the required standard of due diligence has its basis in international law, rather than in national legislation, advising that 'it imposes an obligation on the State of origin to adopt and implement national legislation incorporating accepted international standards'. The State of origin is expected to put in place appropriate 'administrative, financial and monitoring mechanisms', requiring that it should have in place a system for the prior authorisation of relevant activities and that it should play an active role in their regulation.[24] Another aspect of due diligence requires that natural or juridical persons at 'risk of significant transboundary harm' should enjoy access to justice in the State of origin, unless the States concerned have agreed on alternative means of redress,[25] a position reflected in the dispute-settlement provisions of the UN Watercourses Convention.[26]

The commentary also links Draft Article 3 on Prevention to Draft Articles 9 and 10, which require inter-State consultation on the preventive measures to be adopted, having regard to the need to achieve an equitable balancing of the interests of the States concerned. In a manner strongly reminiscent of the factors relevant to the international water law principle

---

[23] International Law Association Study Group (2016), *supra*, n. 17.
[24] International Law Commission Draft Articles on Prevention, *supra*, n. 19, arts 6 and 7.
[25] Draft Article 15 on Prevention, *ibid*.
[26] UN Watercourses Convention, *supra*, n. 1, art 32.

of equitable and reasonable utilisation,[27] Draft Article 10 on Prevention provides an open-ended list of factors relevant to such equitable balancing of States' interests. These include:

- the degree of risk of significant harm (including harm to the environment) and the availability of means for its prevention or minimisation;
- the economic and social importance of the harmful activity in question;
- the extent to which either State might contribute to the costs of prevention;
- the economic viability of the activity in question having regard to the costs of prevention (and the availability of alternatives); and
- the standards of prevention otherwise applied by the affected State.

Highlighting the obvious parallels with the principle of equitable and reasonable utilisation as articulated under the UN Watercourses Convention, the commentary to Draft Article 10 explains that this provision 'draws its inspiration from Article 6 of the Convention on the Law of the Non-navigational Uses of International Watercourses'. It further illustrates the factors set out in Draft Article 10 by referring to corresponding international water law cases[28] and conventions.[29] Of course, by subjecting the duty to prevent transboundary harm under the 2001 Draft Articles on Prevention to an equitable balancing of the interests of the States concerned, the ILC further supports the understanding that the no-harm rule as set out under Article 7 of the UN Watercourses Convention is subordinate to the principle of equitable and reasonable utilisation under Articles 5 and 6.

The commentary to Draft Article 3 further suggests that the duty to prevent only applies to harm that is reasonably foreseeable, stating that 'in general, in the context of prevention, a State of origin does not bear the risk of unforeseeable consequences to States likely to be affected by activ-

---

[27] *Ibid.*, art 6(1).
[28] *Streitsache des Landes Wurttemburg und des Landes Preussen gegen das Land Baden (Wurttemburg and Prussia v Prussia) betreffend die Donauversinkung*, Garman Staatsgerichtshof, 18 June 1927 (*Donauversinkung* case).
[29] 1976 Convention on the Protection of the Rhine against Pollution from Chlorides, (3 December 1976) 16 *International Legal Materials* 265 and 1994 Additional Protocol (25 September 1991) 1994 UNTS 423; 1973 Agreement on the Permanent and Definitive Solution to the International Problem of the Salinity of the Colorado River, (30 August 1973) 12 *International Legal Materials* 1105.

ities within the scope of these articles'. However, the Commission also advises that the obligation to 'take all appropriate measures ... extends to taking appropriate measures to identify activities which involve such a risk'. More generally, as regards a State of origin's duty to ensure that it is adequately informed, the commentary provides that 'due diligence is manifested in reasonable efforts by a State to inform itself of factual and legal components that relate foreseeably to a contemplated procedure and to take appropriate measures, in timely fashion, to address them'. In addition, the commentary invokes the precautionary principle as articulated in Principle 15 of the Rio Declaration, explaining that preventive measures taken under Article 3 'could involve, inter alia, taking such measures as are appropriate by way of abundant caution, even if full scientific certainty does not exist, to avoid or prevent serious or irreversible damage'. In discussing the risk of harm to the environment as a factor involved in the equitable balancing of interests when identifying appropriate preventive measures, the commentary to Draft Article 10 also emphasises the relevance of the precautionary approach. The ILA Study Group on Due Diligence explains, as regards the due diligence conduct of a State of origin under the no-harm rule, that an injured State must demonstrate 'that the State has not put in place the legislative and regulatory framework which would have enabled it to become aware of the risk, to measure its probability and gravity, and to take measures aimed at preventing the harm'.[30] Clearly, measures requiring EIA of potentially harmful projects and activities and facilitating effective ongoing inter-State exchange of relevant water-related data and information have a central role to play in discharging a watercourse State's due diligence obligations under international water law.

In each environmental context in which the concept of due diligence is employed in international law, a number of variable factors may dictate to some extent the standard of care expected of a State of origin. The key factor is that of the degree of risk and hazard involved, to which the degree of care to be exercised should be proportional. According to the International Law Commission commentary to Draft Article 3 on Prevention, 'activities which may be considered ultrahazardous require a much higher standard of care in designing policies and a much higher degree of vigour on the part of the State to enforce them'. The operation of

---

[30] International Law Association Study Group on Due Diligence in International Law, *First Report* (2014).

this factor is elaborated upon somewhat in Draft Article 10(a) and (c) on Prevention, which require, in the identification of appropriate preventive measures, a balancing of the degree of risk of significant transboundary (environmental) harm against the availability of means of preventing or minimising such risk and the possibility of repairing the harm or restoring the environment. The International Law Association Study Group also stresses the relative importance of the hazard involved, concluding:

> In international environmental law, a higher standard of care is required when inherently hazardous activities are undertaken; here, the degree of diligence varies in light of the level of risk. Advances in scientific understanding and technological capabilities can also increase the degree of care required over time.[31]

The other key factor to be taken into account in determining whether a State has exercised adequate due diligence is that of the State's degree of economic development and its related governance and technical capacity.[32] The commentary to Draft Article 3 on Prevention explains that:

> It is, however, understood that the degree of care expected of a State with a well-developed economy and human and material resources and with highly evolved systems and structures of governance is different from States which are not so well placed.

While this approach reflects the principle of common but differentiated responsibilities set out in Principle 3 of the Rio Declaration, it remains clear that an underdeveloped State lacking such capacity is not exempt from its obligations under the no-harm rule. In every case 'vigilance, employment of infrastructure and monitoring of hazardous activities in the territory of the State, which is a natural attribute of any Government, are expected'.[33] The capacity of the State of origin might be particularly relevant in taking appropriate preventive measures on the basis of the precautionary principle. The Seabed Disputes Chamber of the International Tribunal of the Law of the Sea, has confirmed that precautionary measures envisaged under Rio Principle 15 must be applied by States 'according to their capabilities', though it also found that 'the reference to different capabilities in the Rio Declaration does not, however,

---

[31] *Supra*, n. 17.
[32] *Trail Smelter Arbitration (US v Canada)* (1941) 3 RIAA 1911.
[33] International Law Commission Draft Articles on Prevention, *supra*, n. 19.

apply to the obligation to follow "best environmental practices" where these are set out in an applicable measure.[34]

4.2.2.2 Context-specific standards

In its 2011 Advisory Opinion, the ITLOS Seabed Disputes Chamber succinctly outlines the 'variable' character of due diligence obligations, such as the no-harm rule:

> The content of due diligence obligations may not easily be described in precise terms. Among the factors that make such a description difficult is the fact that due diligence is a variable concept. It may change over time as measures considered sufficiently diligent at a certain moment may become not diligent enough in light, for instance, of new scientific or technological knowledge. It may also change in relation to the risks involved in the activity.[35]

Therefore, while the ILC has developed authoritative secondary rules on the scope of a State's international legal obligation to prevent transboundary harm, and of that State's responsibility for breach thereof,[36] 'it is to primary rules of conduct, rather than secondary rules of responsibility, that we must look to determine the applicable standard of behaviour'.[37] Despite understandable concerns regarding normative fragmentation over divergences in the sectoral application of due diligence standards, there is no inherent contradiction between the very general standard of due diligence articulated in the *Corfu Channel* case and the more specific manifestations required in particular sub-branches of international law, such as international environmental or water resources law. In this regard, the commentary to the Draft Articles on Prevention expressly refers to provisions of a number of multilateral environmental agreements (MEAs), from which 'an obligation of due diligence as the standard basis for the protection of the environment from harm can be deduced',

---

[34] International Tribunal of the Law of the Sea (ITLOS) Seabed Dispute Chamber, *Responsibilities and Obligations of States Sponsoring Persons and Entities with Respect to Activities in the Area* (ITLOS Case No 17, Advisory Opinion, 1 February 2011), para 131.
[35] *Ibid.*, para 117.
[36] International Law Commission Draft Articles on Prevention, *supra*, n. 19; International Law Commission, Articles on Responsibility of States for Internationally Wrongful Acts, *Report of the International Law Commission*, Fifty-Third Sess. (2001), UN Doc A/56/10, at 43–59, 43–59.
[37] International Law Association Study Group (2014), *supra*, n. 30.

and any one of which might in the particular circumstances 'constitute a necessary reference point to determine whether measures adopted are suitable'. Creating a clear link to the practice of international water law, MEA provisions expressly listed include Article 2(1) of the 1992 UNECE Water Convention.[38] Indirectly relevant provisions listed include Article 2(1) of the 1991 (Espoo) Convention on Environmental Impact Assessment in a Transboundary Context.[39]

As a particularly environmentally progressive example of the primary rules applying to shared water resources, the 1992 UNECE Water Convention provides some detail regarding appropriate measures 'to prevent, control and reduce transboundary impact'. Article 2 expressly stipulates measures for the control of pollution, for ecologically sound and rational water management, for conservation of water resources, and for conservation and restoration of ecosystems. It further provides that such measures shall be taken at source, where possible, and 'shall not result in a transfer of pollution to other parts of the environment'. In addition, Article 2 directs that, in taking such measures, the Parties shall be guided by the precautionary principle, the polluter pays principle and the principle of inter-generational equity. It requires that the Parties cooperate in this regard 'in order to develop harmonized policies, programmes and strategies covering the relevant catchment areas'. Article 3 goes further still, requiring that 'the Parties shall develop, adopt, implement ... relevant legal, administrative, economic, financial and technical measures' for, inter alia, control of pollution emissions through low- and non-waste technology, licensing and monitoring of point-source waste water discharges, discharge limits based on best available technology, special requirements related to the protection of ecosystems, treatment of municipal waste water, application of best available technology to control nutrient inputs from point and non-point sources, application of environmental impact assessment, protection of groundwaters, and contingency planning for accidental pollution. More generally, it requires the setting of comprehensive emissions limits for discharges and of water-quality objectives and criteria, and also that, in so doing, Parties have regard to those industries or hazardous substances controlled under existing applicable conventions or regulations. Though not quite as detailed, Article

---

[38] *Supra*, n. 2.
[39] 1991 Convention on Environmental Impact Assessment in a Transboundary Context (Espoo), (1991) 30 *International Legal Materials* 802.

21 of the UN Watercourses Convention requires States, in preventing, reducing and controlling pollution of an international watercourse, to take steps to harmonise their policies and to agree joint measures, such as:

(a) Setting joint water quality objectives and criteria;
(b) Establishing techniques and practices to address pollution from point and non-point sources;
(c) Establishing lists of substances the introduction of which into the waters of an international watercourse is to be prohibited, limited, investigated or monitored.

Significantly, Article 20 of the Convention requires that '[w]atercourse States shall, individually and, where appropriate, jointly, protect and preserve the ecosystems of international watercourses'.

While these two globally applicable framework water conventions provide a rich and helpful source of primary rules to inform the due diligence required of watercourses States under the duty of prevention in international water law, it is important to remember that the duty remains one relating to conduct rather than to result, and that certain variable factors might impact upon the standard of conduct expected in the circumstances. The general obligation of prevention as it appears in most environmental conventions does not establish a strict obligation not to pollute (obligation of result), but only an obligation to 'endeavour' under the due diligence rule to prevent, control and reduce pollution (obligation of conduct).

4.2.2.3 Substantive and procedural due diligence

In the *Pulp Mills* case, the Court elaborated upon both the procedural and substantive aspects of the duty of prevention, as well as upon the intricate interrelationship between these aspects. Substantive requirements would include the adoption and effective enforcement of appropriate domestic legal controls on water abstraction or pollution or on the protection of the shared watercourse and its related ecosystems. Procedural due diligence includes the requirements for early notification and consultation and, where necessary, negotiation in respect of potentially harmful planned projects or uses of international water resources. The International Court of Justice highlighted that each of these requirements could only meaningfully be performed in conjunction with an EIA of the likely transboundary effects. The Court regarded procedural cooperation and substantive rules as 'intrinsically linked' functionally, explaining that

> it is by cooperating that the States concerned can jointly manage the risks of damage to the environment that might be created by the plans initiated by one or other of them, so as to prevent the damage in question, through the performance of both procedural and substantive obligations ... whereas the substantive obligations are frequently worded in broad terms, the procedural obligations are narrower and more specific, so as to facilitate the implementation ... [of substantive rules] ... through a process of continuous consultation between the parties concerned.[40]

The Court has confirmed the functional interdependence of the substantive and procedural requirements of international water law in the joined *San Juan River* cases, stating:

> If the environmental impact assessment confirms that there is a risk of significant transboundary harm, the State planning to undertake the activity is required, in conformity with its due diligence obligation, to notify and consult in good faith with the potentially affected State, where that is necessary to determine the appropriate measures to prevent or mitigate that risk.[41]

4.2.2.4 Ecological due diligence

The International Law Association Study Group on Due Diligence points out that 'the content of the obligation may also change in line with scientific and technological advances'.[42] One can clearly observe in the field of international water law notable recent growth in ecological awareness and developments in scientific understanding, alongside a corresponding emphasis in State and treaty practice on the legal obligations of States regarding the protection and preservation of international watercourse ecosystems, and the emergence of sophisticated methodologies that inform the normative implications of such obligations. Such methodologies include increasingly detailed parameters for assessing minimum environmental flows in a shared watercourse, which have already facilitated judicial recognition of a corresponding legal obligation to maintain a minimum environmental flow regime,[43] and the rapidly evolving ecosystem services concept, which focuses on the essential natural services

---

[40] *Pulp Mills* case, *supra*, n. 16, paras 68 and 77.
[41] *Supra*, n. 11, paras 104 and 168.
[42] International Law Association Study Group (2014), *supra*, n. 30.
[43] *Kishenganga Arbitration* case, *supra*, n. 12, para 454.

furnished by functioning riverine ecosystems and provides a methodology for the economic and social valuation of natural ecosystems. The International Court of Justice has recently determined for the first time that 'damage to the environment, and *the consequent impairment or loss of the ability of the environment to provide goods and services,* is compensable under international law' (emphasis added) and proceeded to assign a monetary value in compensation for the loss of four specific classes of ecosystem services.[44]

Ecosystems obligations are not new in international water law. For example, the 1992 UNECE Water Convention expressly requires Parties to apply 'the ecosystems approach'[45] and defines the 'transboundary impact', that the Parties are to 'take all appropriate measures to prevent, control and reduce',[46] to include 'effects on human health and safety, flora, fauna, soil, air, water, climate, landscape … or the interaction among these factors'.[47] However, new scientific and methodological advances will inevitably inform the practical implications of the relevant due diligence requirements. Highlighting the flexibility and adaptability of the due diligence concept, the recent ILA study notes that, 'as international law develops into new, more complex areas … due diligence is increasingly viewed as an important tool in responding to such challenges'.[48] It also notes that '[t]he extent of risk or advances in scientific knowledge that allow us to perceive more accurately the extent of risk (either higher or lower) will also influence the degree of diligence required'. Clearly, methodological developments such as those outlined above can do much to clarify the precise nature of the conduct expected of a State of origin under the duty of prevention in the specific context of international watercourses and beyond.

---

[44] *Certain Activities* case, [2018], *supra*, n. 13, paras 42 and 75–87.
[45] UNECE Water Convention, *supra*, n. 2, art3(1)(i).
[46] *Ibid.*, art 2(1).
[47] *Ibid.*, art 1(2).
[48] International Law Association Study Group (2016), *supra*, n. 17.

4.3 Relationship with equitable and reasonable utilisation

The relationship between the duty to prevent significant transboundary harm and the overarching cardinal principle of equitable and reasonable utilisation has long been an issue of some controversy. However, on careful examination of the 1997 UN Watercourse Convention and its *travaux préparatoires* it quickly becomes apparent that the no-harm rule, and other substantive rules of international water law, such as those relating to environmental protection and the maintenance of ecosystems, are subject to the doctrine of equitable and reasonable utilisation. In other words, the duty to prevent has in principle only a limited, though in practice probably profound, effect on the operation of the balancing of interests required under equitable and reasonable utilisation. In certain circumstances, one watercourse State's use of shared waters would have to be tolerated, even if it caused significant harm to another, where the offending use represented the most equitable and reasonable allocation of benefits taking account of all relevant considerations. This hierarchy of substantive rules is apparent from the wording of Article 7 of the UN Watercourses Convention, which provides that watercourse States whose use causes significant harm 'shall ... take all appropriate measures, *having due regard to the provisions of articles 5 and 6* ... to eliminate or mitigate such harm and, where appropriate, *to discuss the question of compensation*' (emphasis added). This clearly implies that Article 7, setting out the parameters of the no-harm rule, is subordinate to the principle of equitable and reasonable utilisation set out in Article 5 and 6. The International Law Association's update on the 1966 Helsinki Rules,—the 2004 Berlin Rules on Water Resources—inverts this relationship somewhat but endorses the same hierarchy of rules, providing that 'Basin States shall in their respective territories manage the waters of an international drainage basin in an equitable and reasonable manner *having regard for the obligation not to cause significant harm to other basin States*' (emphasis added).[49]

Article 6(1) of the UN Watercourses Convention also list 'the effects of the use or uses of the watercourses in one watercourse State on other

---

[49] ILA, *Report of the Seventy-First Conference of the International Law Association* (2004) art 12(1) ('Berlin Rules').

watercourse States' as one factor among others relevant to the determination of equitable and reasonable utilisation. In the *Gabčíkovo-Nagymaros* case the International Court of Justice strongly endorsed the principle of equitable and reasonable utilisation as the governing rule of international water law and the one on which that dispute should turn.[50] This subordination of the no-harm rule possibly reflects the fact that, in many international watercourses, lower basin States tend to develop earlier and so a strict prohibition on causing significant harm might effectively serve to protect existing rights and would thus impede opportunities for newly developing upstream States to pursue their legitimate interests. This would amount to yet another illustration of the pervasive influence of the concept of 'solidarity' and the principle of 'common but differentiated responsibility' (CBDR). However, continuing disagreement between States regarding the primacy of the equitable and reasonable utilisation principle tends to reflect the well-established fault-line in international water law between upstream and downstream watercourse States.

It would be a mistake, however, to imagine that these two key substantive rules of international water law are often, if at all, likely to come into genuine conflict. Relying on a celebrated formulation of the no-harm rule articulated by the German *Staatsgerichtshof* (State Constitutional Court) in relation to a 1927 complaint taken by the *Länder* of Württemburg and Prussia against Baden (the *Donauversinkung* case), the leading scholar in this field of international law suggests that 'the *sic utere tuo* principle is not only fully compatible with that of equitable utilization, it essentially merges with the latter principle', explaining further that:

> It is thus the flexibility of the no-harm rule that makes it compatible, even if not entirely identical, with the principle of equitable utilization ... rather than prohibiting the causing of harm per se, the law takes into account surrounding circumstances. This same process is followed in arriving at an equitable and reasonable allocation of the uses and benefits of shared freshwater resources ... There is therefore no need to 'reconcile' the no-harm and equitable utilization principles. They are, in reality, two sides of the same coin.[51]

---

[50] *Case Concerning the Gabčíkovo-Nagymaros Project (Hungary/Slovakia)* (1997) *ICJ Reports* 7, paras 78, 85, 147 and 150.
[51] McCaffrey, *supra*, n. 15, at 357 and 370–371.

4.4 Relevance for environmental harm

There can be little doubt that the no-harm rule envisages environmental pollution and ecosystems damage, and resulting loss of beneficial ecosystem services, as centrally relevant classes of harm to be prevented. The inclusion of Article 7 in the UN Watercourses Convention, in conjunction with Articles 20–23, demonstrates that environmental protection featured prominently in the International Law Commission's thinking when developing the Draft Articles which preceded and provided the basis for the Convention. The environmental implications of the no-harm rule must be understood in the light of the rapidly evolving international legal framework for environmental protection, which inevitably informs the normative content of environmental due diligence in international law more generally. In many cases the standard of State conduct required in order to satisfy due diligence may be determined by reference to internationally agreed minimum standards in the environmental field. Such standards might relate, for example, to the conduct of environmental impact assessment of projects likely to impact a shared watercourse[52] or to the management of hazardous chemicals presenting a risk of serious water pollution.[53]

Although environmental damage is not the only, or possibly even the main cause of harm to other watercourse States, it is telling that the UN Watercourses Convention includes dedicated substantive provisions obliging watercourse States to protect the ecosystems of international watercourses, to prevent significant harm resulting from pollution, to protect the marine environment impacted by international watercourses, and to ensure the sustainable development of international watercourses.[54] While their precise normative relevance to the no-harm rule and the principle of equitable and reasonable utilisation remains a little opaque, the express inclusion of such obligations within the text of the Convention reflects normative expectations which must inevitably influence the outcome of any process of balancing the interests of watercourse States, so that pollution of international watercourses and degradation of

---

[52] For example, 1991 Convention on Environmental Impact Assessment in a Transboundary Context (Espoo), (1991) 30 *International Legal Materials* 802.
[53] For example, 2001 (Stockholm) Convention on Persistent Organic Pollutants, (2001) 40 *International Legal Materials* 532.
[54] UN Watercourses Convention, *supra*, n. 1, arts 20–24 and 5(1).

aquatic ecosystems might be regarded as a special form of harm, subject to a somewhat different regime from that applicable to general water resources allocation and utilisation.[55]

4.5 Conclusion

While the flexibility and adaptability required of the duty of prevention, and of the due diligence standard of State conduct which it implies, inevitability result in a measure of uncertainty, it must be recognised that this is a problem that characterises the application of substantive rules right across the broader field of international environmental and natural resources law. In international water law, such normative indeterminacy reflects the fact that no two international river basins are similar. However, the advent of detailed rules and related evaluation and assessment methodologies regarding protection of international watercourse ecosystems, and maintenance of the ecosystem services provided thereby, lends a welcome degree of clarity to the relevant primary rules of international water law. This can only serve to shed light on the due diligence conduct expected of watercourse States under the no-harm rule in international law.

---

[55] McCaffrey, *supra*, n. 15, at 364.

5. Cooperation and procedural rules of international water law

5.1 Introduction

Nobody could doubt the absolutely central role of procedural rules within the cooperative framework currently provided by international water law, as these relatively unambiguous rules require the structured exchange of information which is so vital to any meaningful inter-State engagement. The significance of procedural rules for inter-State cooperation has long enjoyed judicial and arbitral recognition, stretching back to the seminal *Lac Lanoux* case in 1957.[1] In its 2010 judgment in the *Pulp Mills* case the International Court of Justice (ICJ) placed great emphasis on the functional importance of procedural requirements, whether conventional or customary, and did much to clarify the inter-relationship between procedural and substantive rules of international water law.[2] For the purposes of determining a breach of either of the key substantive principles of international water law, that of equitable and reasonable utilisation or that of prevention of significant transboundary harm, the Court identified two categories of due diligence obligation inherent to both—procedural and substantive—and clearly linked the duty to notify, and the closely associated obligation to conduct transboundary environmental impact assessment of planned projects, to satisfaction of the procedural due diligence obligations. Substantive due diligence, on the other hand, might for example involve the adoption and enforcement of national legal requirements regarding the management of water pollution. In addition to their role in informing compliance with the substantive obligations, the Court

---

[1] *Lac Lanoux Arbitration (Spain v France)* (Award on 16 November 1957) (1961) 24 *International Law Reports* 101.
[2] *Pulp Mills on the River Uruguay (Argentina v Uruguay)* (Judgment) [2010] *ICJ Reports* 14.

found that the procedural duties of international water law also create binding obligations in their own right, though, by failing to impose an onerous remedy, it suggested that breach of such obligations might not be considered very serious in the absence of material transboundary harm.

The Court highlighted the cascading nature of the procedural obligations flowing from the general duty of States to cooperate which, in the context of planned projects or activities likely to impact adversely upon co-riparian States, requires prior notification and, where necessary, follow-on consultation and negotiation. In order to be adequate, such notification in turn requires some form of environmental impact assessment that takes full account of the transboundary effects of the project or activity in question. Of course, the requirement to conduct transboundary environmental impact assessment of planned projects, now recognised as a general requirement under customary international law, has a parallel in the requirements relating to the ongoing exchange of information on existing water resources utilisation and its environmental impacts, as suggested by the Court's consistent endorsement of a requirement for continuing environmental impact assessment of potentially harmful projects during their operational lifetime.[3] Inevitably, procedural obligations are very closely linked to the establishment of cooperative institutional mechanisms, through which formal exchange of information and inter-State dialogue can take place, and by means of which detailed basin-level procedural rules on such exchange can be further elaborated and implemented. It is hardly coincidental that the rise to prominence of procedural rules of international water law, acknowledged by the ICJ in recent times, has been accompanied by a proliferation of river basin organisations, boundary waters commissions and similar inter-governmental institutions having clear responsibility for cooperative management of shared water resources. This 'institutionalisation' of inter-State cooperation over shared water resources can be understood as reflecting a broader trend in international environmental and natural resources law involving the

---

[3] See *Case Concerning the Gabčíkovo-Nagymaros Project (Hungary/Slovakia)* (1997) *ICJ Reports* 7, Separate Opinion of Judge Weeramantry. See also *Nuclear Tests* case *(New Zealand v France)* (1995) (Request for an Examination of the Situation in Accordance with Paragraph 63 of the Court, Judgment of 20 December 1974) *ICJ Reports* 457; Legality of the Use by a State of Nuclear Weapons in Armed Conflict (Advisory Opinion) (1996) *ICJ Reports* 66.

transition from an international law of coexistence to an international law of cooperation.

The precise nature and application of the judicially recognised obligation in general international law to conduct transboundary EIA in respect of potentially harmful projects or activities remains, however, in need of some clarification. Despite the Court's emphatic emphasis in the *Pulp Mills* case on the requirement to notify a co-riparian State as soon as possible of a project or activity with potential transboundary effects, in a subsequent case involving Nicaragua and Costa Rica the ICJ found that EIA precedes and informs the need for notification, which only becomes necessary where 'the environmental impact assessment confirms that there is a risk of significant transboundary harm'.[4] It further found in that case that, 'in light of the absence of risk of significant harm, Nicaragua was not required to carry out an environmental impact assessment'. At the same time, in a joined case, the Court observed that 'to conduct a preliminary assessment of the risk posed by an activity is one of the ways in which a State can ascertain whether the proposed activity carries a risk of significant transboundary harm', thereby requiring an environmental impact assessment.[5] It seems, therefore, that the Court has retreated from its earlier finding, in *Pulp Mills*, that a State must inform 'as soon as it is in possession of a plan which is sufficiently developed to permit a preliminary assessment ... [or] ... at the stage when the relevant authority has had the project referred to it with the aim of obtaining initial environmental authorization and before the granting of that authorization'. This position, requiring early notification in advance of EIA, if maintained, could facilitate the early and constructive engagement of States likely to be impacted by planned measures in the framing and conduct of the EIA study. For example, such States might be given an opportunity to comment on the terms of reference for an EIA study, and thus on its material scope and methodological approach, in advance of formal notification. This might allow watercourse States to pre-empt and address controversial issues arising in a timely manner and, thereby, to avoid project delays and extended water-related transboundary disputes.

---

[4] *Certain Activities Carried Out by Nicaragua in the Border Area (Costa Rica v Nicaragua)* and *Construction of a Road Along the San Juan River (Nicaragua v Costa Rica)* (Judgment on 16 December 2015) (2015) *ICJ Reports* 665.

[5] *Construction of a Road* case.

It is nonetheless clear that inter-State notification on the basis of an EIA study plays a key role in ensuring that environmental (and, increasingly, social) considerations relating to a planned or continuing use of an international watercourse are adequately understood and communicated. Such considerations may therefore be taken properly into account, either as a factor within the balancing process that lies at the heart of equitable and reasonable utilisation, or as a key component of the procedural due diligence element of the duty to prevent significant transboundary harm. In either role transboundary EIA may also facilitate application of certain associated principles and approaches of international water and environmental law, such as the precautionary principle, thereby allowing these concepts to inform the actions of national decision-makers. However, as a 'front-loaded' assessment mechanism normally employed to inform one-time decision-making processes determining whether to permit implementation of a planned project, use or activity, it is not at all clear that EIA, at least as currently employed and conducted, is suited to facilitating the kind of flexible adaptive management approach which is increasingly regarded as suitable for complex ecological systems such as major watercourses. Though watercourse States are under an ongoing obligation to exchange information on the conditions in the watercourse, this is neither sufficiently elaborated in conventional instruments nor developed in general international practice to ensure effective ecosystems protection.

5.2 The general duty of cooperation

The general obligation of States to cooperate in the resolution of international problems is widely accepted and receives support from as authoritative a legal source as the UN Charter,[6] and has long been supported in global declarative practice.[7] In the field of international water resources, it is largely given practical effect by means of various associated rules of

---

[6] Charter of the United Nations (adopted 26 June 1954, entered into force 24 October 1945) 1 UNTS 16, art 1(3).

[7] For example, UN General Assembly, Declaration on Principles of International Law concerning Friendly Relations and Co-operation among States in accordance with the Charter of the United Nations (24 October 1970) UN Doc A/RES/2625 (XXV).

procedural conduct rapidly emerging as international custom, including the duties to notify of planned measures, to consult, to negotiate and to warn, as well as duties relating to the ongoing exchange of relevant data and information. Whatever the precise legal status in customary international law of the general duty to cooperate, it appears more firmly established and highly developed in its application to the utilisation and environmental protection of shared natural resources, and in particular water resources, where the requirements of notification and prior consultation regarding planned projects, based on an adequate environmental and social impact assessment, are universally understood as ancillary to the equitable and reasonable utilisation of a shared resource and to the prevention of significant transboundary harm. This is supported by numerous declarations and recommendations referring to the duty and elaborating some means for its implementation,[8] as well as the many international watercourse agreements expressly alluding to the obligation to cooperate.[9]

The Tribunal in the 1957 *Lac Lanoux* arbitration emphatically recognised the duty of States to cooperate in the use of the waters of an international watercourse, identifying good faith cooperation in the conclusion of international agreements as the key means of ensuring the prevention of transboundary harm.[10] More recently, the judgment of the International Court of Justice in the *Gabčíkovo-Nagymaros* case reflected the procedural obligation to cooperate,[11] which has received similar support from international tribunals in a range of environmental disputes.[12] While Article 8 of the UN Watercourses Convention specifically addresses the 'general obligation to cooperate', Part III of the Convention, comprising Articles

---

[8] Including UNCHE, *Report of the UN Conference on the Human Environment* (5–16 June 1972) UN Doc A/CONF.48/14/Rev.1, pt I, ch I, (reprinted in 11 *International Legal Materials* 1416), Principle 24; UNCED, Rio Declaration on Environment and Development (1992) (reprinted in 31 *International Legal Materials* 876) UN Doc A/CONF/151/5 Rev.1, Principle 19.

[9] Early examples include the Berne Convention on the International Commission for the Protection of the Rhine (adopted 29 April 1963, entered into force 1 May 1965) 994 UNTS 3; Great Lakes Water Quality Agreement between Canada and the United States (adopted and entered into force 22 November 1978) 30 UST 1383, arts 7–10.

[10] *Lac Lanoux Arbitration, supra*, n. 1.

[11] *Supra*, n. 3, para 17.

[12] For example, *MOX Plant* case *(Ireland v United Kingdom)*, Provisional Measures (Order) [2001] *ITLOS Reports* 10, para 89.

11–19, contains detailed procedural rules requiring watercourse States to notify, consult and negotiate in relation to planned measures which may have adverse effects on other watercourse States. The International Law Association's 2004 Berlin Rules contain a Chapter XI dedicated to 'International Cooperation and Administration' and setting out detailed rules on, inter alia, exchange of information, notification of programmes, plans, projects or activities, and consultation, while the commentary thereto asserts that 'the duty of cooperation is the most basic principle underlying international water law'.[13] It is clear that permanent river basin organisations play a key role in facilitating the intense procedural engagement required under the duty to cooperate, and the UN Watercourses Convention expressly encourages watercourse States to enter into institutional arrangements to facilitate inter-State cooperation.[14] While States cannot generally be compelled to establish or join common management institutions, a watercourse State's bona fide participation in such arrangements may help to demonstrate satisfaction of the procedural obligations inherent to cooperation.

The duty to cooperate can therefore be understood as a composite obligation, largely consisting of a range of procedural requirements, any one or more of which might be applicable in a given situation. Due to their nature, such procedural rules are usually understood to embody normatively clear and unambiguous State obligations. However, while any violation of the duty to cooperate, or of any of its related procedural obligations, will amount to an internationally wrongful act per se, any redress awarded in international litigation is unlikely to be significant where this is not accompanied by material harm to the State complaining of the breach.[15] Where material harm has actually been caused to another watercourse State, breach of any applicable procedural requirement is likely to amount to a failure to meet the standards of procedural due dili-

---

[13] International Law Association, 'Berlin Rules on Water Resources Law', in *Report of the Seventy-First Conference of the International Law Association* (ILA, Berlin, 2004), commentary to art 11, at 20.
[14] Most notably in the 1997 UN Convention on the Law of the Non-Navigational Uses of International Watercourses (adopted 21 May 1997, entered into force 17 August 2014), (1997) 36 *International Legal Materials* 700, art 8(2).
[15] See *Pulp Mills* case, *supra*, n. 2, paras 79, 121–122 and 275–276; *Certain Activities* and *Construction of a Road* cases, *supra*, nn. 4 and 5, paras 224–226.

gence necessary to discharge the duty to prevent significant transboundary harm and, in such a case, more meaningful redress may be available.[16]

5.3 Procedural rules

The individual procedural requirements arising under international water law are linked in a cascading suite of closely inter-connected rules, together comprising a continuum of inter-State engagement. For example, notification of a planned measure or routine exchange of information regarding the state of a shared river might give rise to inter-State consultation and might, if necessary, lead to further inter-State negotiation with a view to resolving differences arising. Indeed, the ICJ observed in the *Pulp Mills* case that the procedural obligations of international water law together form 'an integrated and indivisible whole'.

5.3.1 Duty to Notify

As a matter of practice, States have long tended to provide neighbouring States with prior notice of plans to exploit a shared natural resource and it is generally agreed that this amounts to an obligatory requirement under customary international law or as a principle generally recognised in international environmental law. Several States have long relied on the duty to provide prior notification in the course of international disputes[17] and it receives broad support in important recent conventional and declaratory instruments, including, inter alia, the 1992 Convention on Biological Diversity[18] and the Rio Declaration, Principle 19 of which provides that:

> States shall provide prior and timely notification and relevant information to potentially affected States on activities that may have a significant adverse

---

[16] See *Certain Activities* case, *supra*, n. 4, paras 139 and 142, though responsibility in this case was established on the basis of Nicaragua's unlawful violation of Costa Rica's territorial sovereignty, *ibid.*, para 99.
[17] *Lac Lanoux Arbitration*, *supra*, n. 1.
[18] United Nations Convention on Biological Diversity, (1992) 31 *International Legal Materials* 822, art 14(d).

transboundary environmental effect and shall consult with those States at an early stage and in good faith.[19]

A number of issues arise in relation to the general duty to notify. For example, there is often no clear guidance to inform determination of which States are likely to be affected by a particular activity and consequently entitled to notification. Few treaties requiring notification provide detail on this question, though they may provide for the elaboration of further guidance on this and other matters by treaty bodies or secretariats. The International Law Commission (ILC) has suggested, in several different contexts, that where a number of States are likely to be affected, notification and consultations should take place within the framework of an international organisation. This normative uncertainty ought not to cause very significant problems in the context of international watercourses where the mutual interdependence of riparians is clearly understood, and where many international basins are served by river basin organisations (RBOs) or other institutional bodies of one form or another. Where a watercourse involves a large number of riparian States the issue of notification highlights the potential benefits of a common management approach with multilateral institutions to facilitate routine notifications. Also, there may be difficulty in identifying the types of activities and forms of injury which the State of origin must notify to the potentially affected States. This may depend on the nature of the shared water resource being utilised or the potential scale or probability of harm to which any activity may give rise. Once again, the ILC has observed that whether or not an activity requires notification should be determined on an ad hoc basis, having regard to all the surrounding circumstances. However, it is ultimately left to the notifying State to assess whether there is likely to be a significant adverse effect and to decide which other watercourse States may be among those so affected.

It appears that the duty requires the State of origin to notify promptly, usually no later than when informing its own public, and certainly before the commencement of the activity in question or before any related construction work is authorised. The notification should indicate a reasonable timeframe within which a response is required and the source State may also require the affected State to supply reasonably obtainable infor-

---

[19] Rio Declaration on Environment and Development, (1992) 31 *International Legal Materials* 874, UN Doc A/CONF.151/26 (vol 1).

mation relating to the likely impacts within the territory of the affected State. Failure to respond to notification may give rise to a presumption of acquiescence in the proposed activity and potentially preclude any subsequent assertion that the source State failed to take its interests into account. Notification is subject to the general requirement that States perform their international obligations in good faith, potentially informing, for example, how early any notification ought to be or a reasonable period within which a response ought to be delivered. The requirement of good faith in the conduct of international relations also suggests that, pending the reply from the notified State, the source State is effectively debarred from implementing its plan.

The duty to notify is included in all significant water-related agreements. The 1992 UNECE Water Convention requires parties to enter into bilateral or multilateral agreements or other arrangements which provide for the establishment of joint bodies to have responsibility for, inter alia, 'the exchange of information on existing and planned uses of water and related installations that are likely to cause transboundary impact' and to 'participate in the implementation of environmental impact assessments relating to transboundary waters, in accordance with appropriate international regulations'. Probably the most developed conventional articulation of the duty to notify is set out under Part III of the 1997 UN Watercourses Convention. Comprising Articles 11–19, Part III relates to 'Planned Measures' and contains interlinked procedural rules requiring watercourse States to notify, consult and negotiate in relation to planned measures which may have adverse effects. Article 12 of the Watercourses Convention provides that:

> Before a watercourse State implements or permits the implementation of planned measures which may have a significant adverse effect upon other watercourse States, it shall provide those States with timely notification thereof. Such notification shall be accompanied by available technical data and information, including the results of any environmental impact assessment, in order to enable the notified States to evaluate the possible effects of the planned measures.

Some uncertainty arises over whether the term 'significant adverse effect upon other watercourse States', as employed in Article 12 of the Convention, could include adverse effects on possible future developments in those watercourse States, or is limited to adverse effects on the existing state of affairs. As future uses, at least those that are not at a reasonably advanced stage of planning, cannot easily be assessed with any precision, it is often not practical or reasonable to delay one State's planned measures on the ground that a development may take place in another State at some unspecified time in the future. However, Article 17(2) expressly requires that consultations and negotiations concerning planned measures 'shall be conducted on the basis that each State must in good faith pay reasonable regard to the rights and legitimate interests of the other State', which might obviously include future utilisation of the waters within its territory. Also, if future uses are not considered, then only existing uses would enjoy procedural protection, thereby favouring and encouraging immediate, even premature development of such uses. It is significant that future uses enjoy some substantive protection under the principle of equitable and reasonable utilisation as formulated under Article 5 and 6, with the 'existing and potential uses of the watercourse' expressly listed among the factors required to be taken into account as relevant to equitable and reasonable utilisation.

Where the watercourse State planning measures decides not to notify another watercourse State, the latter State may, where it 'has reasonable grounds to believe' that the planned measures may have a significant adverse effect upon it, request notification under Article 18 and must accompany such a request with a documented explanation setting forth its grounds. Continued disagreement on the issue requires the States concerned to enter into consultations and negotiations as provided for under Article 17 of the Convention during which the State planning the measures shall, if requested, 'refrain from implementing or permitting the implementation of those measures for a period of six months unless otherwise agreed'. The obligation to refrain may be waived, however, subject to certain procedures, 'in the event that the implementation of planned measures is of the utmost urgency in order to protect public health, public safety or other equally important interests'. Similarly, the obligation may be substantially relaxed in an 'emergency' situation, where a watercourse State may 'immediately take all practical measures necessitated by the circumstances to prevent, mitigate and eliminate harmful effects of the emergency'. In such a situation, a watercourse State is required to, 'without delay and by the most expeditious means available, notify other potentially affected States and competent international organizations of any emergency originating within its territory'.

The Watercourses Convention provides further details in relation to the duty to notify other watercourse States of planned measures. Article 13 provides that, unless otherwise agreed, the notifying State shall allow notified States a period of six months within which to study and evaluate the measures and to communicate their findings. This period must be extended for a further six months at the request of a notified State 'for which the evaluation of the planned measures poses special difficulty'. Furthermore, Article 14 requires that during this period the notifying State shall provide on request 'any additional data and information that is available and necessary for an accurate evaluation'. As only 'available' technical and additional data and information must be provided, the notifying State may be discouraged from generating such data and information during the planning of the measures. However, it is significant that Article 12 of the Convention expressly requires that the results of any environmental impact assessment accompany the notification. Significantly, Article 14 also stipulates that, during the six-month period referred to in Article 13, the notifying State 'shall not implement or permit the implementation of the planned measures without the consent of the notified States'. Once again, the obligation to refrain is waived in cases of the utmost urgency and relaxed in the case of emergency situations.

Notified States are required, under Article 15, to communicate their findings as early as possible during this period and, where a State finds that the planned measures would be inconsistent with the principle of equitable and reasonable utilisation or with the obligation not to cause significant harm, to document the reasons for the finding. The main aim of the notification procedure is thus twofold: to facilitate agreed, practical implementation of the principle of equitable and reasonable utilisation contained in Article 5, and to ensure compliance with the duty not to cause significant harm contained in Article 7. Where a notifying State receives no reply within the applicable period, it may, under Article 16, 'proceed with the implementation of the planned measures, in accordance with the notification and any other data and information provided to the notified States'. Moreover, Article 16 expressly provides that any subsequent claim to compensation by a notified State which has failed to reply within the applicable period may be offset by the costs incurred by the notifying State for action undertaken after the expiration of that period. However, failure to reply cannot automatically be construed as consent and a notifying watercourse State remains bound by the principle of equitable and reasonable utilisation and by the obligation to prevent significant harm irrespective of a notified State's failure to reply to a notification.

Though the UN Watercourses Convention provides some detail regarding the notification process, the general provisions contained in both the 1992 and 1997 global framework conventions merely provide a framework within which individual States sharing a watercourse can develop more specific regimes of inter-State procedural interaction to meet the particular needs and characteristics of that watercourse.

5.3.2　Ongoing Exchange of Information

In addition to requirements to notify neighbouring States of planned activities or of particular incidents which might give rise to transboundary harm, it is increasingly common for water-related treaties to require State parties to engage in the continuous monitoring of water utilisation and of sources and effects of pollutants, and to exchange information so collected with the other riparian States, often by means of international institutions responsible for the implementation of the treaty regime in question. Many bilateral and multilateral water agreements make express provision for the regular exchange of information, with notable examples including Article 24(C) of the 1995 Mekong Agreement,[20] Article 6 of the 1960 Indus Waters Treaty[21] and Article 3(6) of the Revised 2000 SADC Protocol on Shared Watercourses.[22] Some treaties establish joint bodies or commissions for the collection and exchange of data and information, others permit technical experts from one State to have access to the territory of the other for the purposes of making observations and collecting information, and yet others provide for the establishment of observation stations by one State at the request of another. The 1992 UNECE Water Convention contains very detailed obligations on the exchange of environmental information, with Article 13 requiring the riparian parties to exchange reasonably available data, inter alia, on: environmental conditions of transboundary waters; experience gained in the application and operation of best available technology and results of research and devel-

---

[20] Mekong Agreement, (1995) 34 *International Legal Materials* 864.
[21] Indus Waters Treaty, (1960) *Legislative Texts*, No 98, at 300.
[22] Protocol on Shared Watercourse Systems in the Southern African Development Community (SADC) Region, (2001) 40 *International Legal Materials* 321.

opment; emission and monitoring data; measures taken and planned to be taken to prevent, control and reduce transboundary impact; and permits or regulations for waste-water discharges issued by the competent authority or appropriate body. Article 13 further requires, inter alia, that the parties 'undertake the exchange of information on their national regulations' and 'facilitate the exchange of best available technology'. Similarly, Article 9(1) of the 1997 UN Watercourses Convention requires, in the context of a general obligation to cooperate, that

> watercourse States shall on a regular basis exchange readily available data and information on the condition of the watercourse, in particular that of a hydrological, meteorological, hydrogeological and ecological nature and related to the water quality as well as related forecasts.

Both global Conventions require that a watercourse State employ its best efforts to provide data or information that is not readily available where requested by another watercourse State, though it may require the payment of reasonable costs incurred. They further require that information made available to other watercourse States be useful and comprehensible, with Article 9(3) of the UN Watercourses Convention requiring States to 'process data and information in a manner which facilitates its utilization by the other watercourse States to which it is communicated'.

This obligation can clearly be understood as an integral element of the obligations of equitable and reasonable utilisation and prevention of significant harm, as it will require each State to provide information pertaining to those factors which must be considered in order to ensure equitable use and prevention of harm. However, there may often exist problems in relation to the effective enforcement of this obligation, at least where the relevant treaty regime does not provide for external review or scrutiny of its performance. At the same time, should damage occur, failure to supply such relevant information may be taken as evidence that the State on whom the duty is incumbent has not exercised due diligence over activities under its jurisdiction and control. Thus, the duty to cooperate in the exchange of information, as with other aspects of the general duty to cooperate, might be germane to any determination of breach of the obligation to prevent transboundary harm. In addition, at the level of inter-State practice, the most significant benefit of regular exchange of information and data tends to be improved relations and closer cooperation between riparian States.

5.3.3 Duty to Consult/Negotiate in Good Faith

The vast majority of water-related treaty instruments also require States to enter into consultations and/or negotiations with a view to reaching agreement on the accommodation of the interests of both the State of origin and the notified States. Such agreement would often involve, for example, strategies for minimising the adverse effects of the relevant water use or planned measure or for compensating the affected State. Even where a potentially affected State has not received notification of planned measures or information regarding continuing uses, it will often be expressly entitled to insist on consultations as soon as it becomes aware of the contested use or proposed activity. However, the duty to enter into consultations can never amount to a requirement to obtain the consent of the objecting State, but merely requires the source State to try to accommodate the legitimate interests and concerns of the objecting State. In the *Lac Lanoux Arbitration*, the Arbitral Tribunal held that to find otherwise would effectively amount to the grant of a veto to the objecting State, which would be an intolerable interference with the sovereignty of the source State. Similarly, in the *Nuclear Tests* cases, the ICJ rejected the Australian argument that States had a right to veto the planned activities of other States on the ground that they posed an unacceptable risk of harm.[23] However, though the source State may proceed irrespective of the outcome of negotiations, the conduct and material contents of such negotiations may prove relevant in any subsequent proceedings relating to its responsibility for damage caused.

In the *Lac Lanoux Arbitration*, the Arbitral Tribunal appears to have found that a customary rule exists establishing the obligation of States to negotiate in relation to the utilisation of international watercourses. The ICJ has considered the obligation to negotiate on a number of occasions and has concluded that it is a fundamental principle at the basis of all international relations.[24] Other judicial statements support the argument that '[t]he obligation to negotiate is a principle of general international

---

[23] *Nuclear Tests case (New Zealand v France)* Interim Protection Order of 22 June 1973, (1973) *ICJ Reports*, at 132; Judgment of Judge de Castro, (1974) *ICJ Reports*, at 368–390.

[24] *North Sea Continental Shelf* cases *(Federal Republic of Germany v Denmark and Federal Republic of Germany v Netherlands)* (1969) *ICJ Reports* 2, at 47.

law'.[25] As noted above, the obligation does not require that agreement must be reached but only that the parties must conduct such negotiation in good faith. Negotiations would not be so conducted where one party terminates the negotiations without justification, imposes unreasonable delays or time limits, fails to adhere to the agreed procedure, or systematically refuses to consider the proposals or the interests of the other party. The Tribunal in *Lac Lanoux* emphasised that the discussions must have a substantive content and must not be limited to purely formal requirements such as taking note of complaints. The ICJ has adopted the position that States must conduct meaningful negotiations, which does not occur where one party limits itself to reiterating its position without considering the possibility of a modification. In the light of this arbitral and judicial guidance, one might define the duty to consult to mean that each State must engage in an exchange of views allowing full consideration of their respective interests in the final determination of the issue in question. In *Lac Lanoux*, the Arbitral Tribunal suggested that any failure to negotiate in accordance with the tenets of good faith may attract the responsibility of the State in breach.

Modern treaty practice in relation to international water resources supports the view that there exists under customary international law an obligation to negotiate in good faith. For example, Article 10 of the 1992 UNECE Water Convention provides that 'consultations shall be held between the Riparian Parties on the basis of reciprocity, good faith and good-neighbourliness, at the request of any such Party'. All recent attempts by international bodies to codify the principal customary rules applying to the utilisation and protection of international freshwaters have emphasised the role of the related duties to consult and negotiate in the implementation of the general duty to cooperate. In the course of its work on the non-navigational uses of international watercourses, the International Law Commission attached very considerable importance to the role of procedural safeguards in achieving equitable sharing of uses and benefits, the prevention of transboundary harm, and effective environment management of shared watercourses. Indeed, Judge Schwebel, the ILC's second special rapporteur on the topic, concluded that the procedural requirements of notification and consultation constituted the

---

[25] *Fisheries Jurisdiction Case (United Kingdom v Iceland)* (1973) *ICJ Reports* 2, at 46 (Dissenting Opinion of Judge Padilla Nervo).

'indispensable minima' expected of a State under general international law.

Article 8 of the 1997 UN Watercourses Convention imposes a general obligation on Watercourse States to cooperate 'on the basis of sovereign equality, territorial integrity, mutual benefit and good faith in order to attain optimal utilization and adequate protection of an international watercourse' and goes on to suggest the establishment of joint mechanisms or commissions to facilitate such cooperation. Article 11 further requires that watercourse States 'shall exchange information and consult each other and, if necessary, negotiate on the possible effects of planned measures on the condition of an international watercourse'. More specifically, Article 17 provides that, where a notified State communicates its objection to planned measures pursuant to Article 15, both States 'shall enter into consultations and, if necessary, negotiations with a view to arriving at an equitable resolution of the situation'. It further requires that 'the consultations and negotiations shall be conducted on the basis that each State must in good faith pay reasonable regard to the rights and legitimate interests of the other State'. Article 18, which sets down the procedures to apply in the absence of notification, requires that the State planning measures and the State objecting in the belief that such measures may have a significant adverse effect upon it shall 'promptly enter into consultations and negotiations in the manner indicated in paragraphs 1 and 2 of Article 17'. Similarly, Article 19, which permits the immediate implementation of planned measures in situations of the utmost urgency, requires the implementing State to promptly enter into such consultations and negotiations. The 1997 Convention would appear to envisage consultation as a step that precedes formal negotiations. For example, Article 3(5) requires that 'States shall consult with a view to negotiating in good faith' while Article 17(1)provides that the States concerned are to 'enter into consultations and, if necessary, negotiations with a view to arriving at an equitable resolution of the situation'. Therefore, consultations need not be adversarial in nature and might consist merely of discussions where information is exchanged in relation to factual matters or the interests or positions of the States in question.

It is interesting to note that, though the 1997 Convention does not require that States adopt watercourse agreements, joint management mechanisms or lists of substances whose input into a watercourse are to be prohibited, limited or monitored, it does dictate that, in certain circumstances, they consult and negotiate in good faith with a view to adopting such mechanisms. More significantly, Article 6(2) requires that 'in the application of Article 5 or paragraph 1 of this article [the factors relevant to the principle of equitable and reasonable utilisation], watercourse States shall, when the need arises, enter into consultations in a spirit of cooperation'. Likewise, Article 7(2) requires a State whose use causes significant harm to consult with the affected State concerning the elimination or mitigation of the harm and, where appropriate, to discuss compensation. Thus, the duty to consult and negotiate in good faith permeates the entire fabric of the Convention and plays an absolutely central role in the effective implementation of the key substantive principles and rules of international water law.

During the course of consultations and negotiations entered into pursuant to Articles 17 or 18 of the UN Watercourses Convention, the State planning the controversial measures in question, if requested by the objecting State, shall 'refrain from implementing or permitting the implementation of the planned measures for a period of six months unless otherwise agreed'. Where no consensus is reached during such consultations and negotiations, Article 33 contains detailed provisions in accordance with which watercourse State parties must seek the peaceful settlement of any dispute concerning the interpretation or application of the Convention.

5.3.4 Duty to Warn

It is increasingly clear that a duty to warn co-riparian States in cases of transfrontier water-related emergency has either become clearly established in customary international law or is in the process of emerging. The ICJ has confirmed the existence of a customary duty to warn in general international law in both the *Corfu Channel*[26] and *Nicaragua*[27] cases. Such a duty, which is intended to give States potentially affected by a catastrophic incident an opportunity to adopt evacuation or other mitigating strategies, is contained in a wide range of environmental

---

[26] *Corfu Channel (United Kingdom v Albania)* (1949) *ICJ Reports*, at 22.
[27] *Nicaragua v United States (Merits)* (1986) *ICJ Reports* 4, at 112, para 215.

treaty instruments.[28] Practice surrounding the duty to warn in customary international law suggests that it is more developed in situations involving the utilisation of a shared natural resource, such as an international watercourse system.

Reflecting international practice, the International Law Association's 2004 Berlin Rules on Water Resources contain an entire Chapter VII on 'Extreme Situations', which contains detailed rules on the duty of States to notify other affected States and competent international organisations in the case of 'Extreme Conditions' or 'Polluting Accidents' and to communicate any events likely to create floods and conditions which meet agreed criteria in relation to the risk of droughts. Article 14 of the 1992 UNECE Water Convention provides that 'the Riparian Parties shall without delay inform each other about any critical situation that may have transboundary impact' and further directs that they 'shall set up, where appropriate, and operate coordinated or joint communication, warning and alarm systems with the aim of obtaining and transmitting information'. In addition, it lists among the tasks to be performed by the joint bodies to be established by the riparian parties 'to establish warning and alarm procedures'.

It remains unclear whether, in transfrontier environmental or water-related emergencies, customary international law imposes any further duties beyond the core duty to warn promptly any States likely to be affected. One could argue that, in order to have practical effect, the duty to warn might be supplemented by consequential duties, such as the duty to provide assistance, the duty to develop contingency plans or the duty to provide such relevant information as would assist the affected States in minimising harmful transboundary effects.

In its work relevant to this area, the International Law Association has alluded to duties which are additional to the primary duty to warn, though it is not at all clear that there exists a general legal obligation to discharge such additional duties.

---

[28] For example, art 6 of the 1989 Convention on the Control of Transboundary Movement of Hazardous Wastes and their Disposal, (1989) 28 *International Legal Materials* 657; art 10 of the 1992 Convention on the Transboundary Effects of Industrial Accidents, (1992) 31 *International Legal Materials* 1333.

In line with developments in customary international law, the 1997 UN Watercourses Convention provides that 'a watercourse State shall, without delay and by the most expeditious means available, notify other potentially affected States and competent international organizations of any emergency originating within its territory'. The notion of an 'emergency' is broadly defined and includes

> a situation that causes, or poses an imminent threat of causing, serious harm to watercourse States or other States and that results suddenly from natural causes, such as floods, the breaking up of ice, landslides or earthquakes, or from human conduct, such as industrial accidents.

The Convention further requires that:

> A watercourse State within whose territory an emergency originates shall, in cooperation with potentially affected States and, where appropriate, competent international organizations, immediately take all practicable measures necessitated by the circumstances to prevent, mitigate and eliminate harmful effects of the emergency.

The express requirement to take 'all practicable measures' could in many situations amount to a duty to provide assistance to affected States or to provide all available further relevant information. The Convention further expressly provides that 'when necessary, watercourse States shall jointly develop contingency plans for responding to emergencies, in cooperation, where appropriate, with other potentially affected States and competent international organizations'. Thus the Convention provide some clarity regarding the likely implications of the duty to warn in this particular field.

5.4 Conclusion

One cannot overstate the systemic importance of the general duty of cooperation and its related rules of procedural inter-State engagement to the effective operation of the rules of international water law, whether conventional or customary in origin. If the principal normative objective of international water law has traditionally concerned balancing the legitimate interests of co-riparian States in their use of a shared watercourse, it follows that those States must communicate to each other their current and planned water uses, their water-related needs and their current and emerging concerns related thereto. Such communication between States requires clear procedures and timelines, along with generally accepted methodologies for assessing impacts, benefits and needs. Experience suggests that only institutionalised cooperation can effectively deliver such a cooperative framework. Thus, the procedural rules of international law provide the framework within which watercourse States can, in a spirit of good faith cooperation, elaborate the necessary procedures and methodological approaches. This in turn allows such States to build mutual trust, leading to yet further enhanced procedural cooperation and, ultimately, to optimised benefits and sustainable management of the shared resource.

Effective inter-State cooperation by means of institutionalised procedural engagement is only likely to become even more important in the light of the challenges facing international water law today. The interrelated global crises appearing on the horizon regarding water resources availability and over-utilisation, climate change and biodiversity loss will demand intense inter-State cooperation over transboundary waters, facilitated by robust institutional mechanisms, in order to ensure the most efficient, beneficial and equitable use of an increasingly scare and valuable shared resource. The focus of international water resources management will increasingly be upon the more urgent objective of 'attaining optimal and sustainable utilization thereof and benefits therefrom', as anticipated presciently under the UN Watercourses Convention.[29]

---

[29] UN Watercourses Convention, art 6(1).

6. Environmental protection and ecosystems conservation

6.1 Introduction

Requirements related to environmental and ecosystems protection have long been integral to the very definition and fabric of international water law, which has primarily been concerned with inter-State cooperation over utilisation and *protection* of shared international watercourses. This is hardly surprising as the environmental status of water, shared or otherwise, would very often dictate the economic or social uses to which it might be applied. Equally, degradation of the ecosystems of shared watercourses might equally result in various types of detrimental impacts upon the interests of co-riparian States. Thus, it has long been clearly understood that environmental and ecological protection is inextricably linked to fulfilment of the twin key substantive obligations of international water law, i.e., the duty to use an international watercourse in an equitable and reasonable manner and the duty to prevent the causing of significant harm to other watercourse States or to the watercourse itself. It is telling, for example, that the International Law Association's (ILA) seminally important 1966 Helsinki Rules, the first comprehensive codification of the emerging and established rules of modern international water law, included an entire Chapter 3 on 'Pollution'.[1] The central importance of environmental protection within international water law is further confirmed by the fact that almost every major inter-State dispute over shared waters to have given rise to judicial or arbitral resolution in recent years

---

[1] International Law Association, Helsinki Rules on the Uses of the Waters of International Rivers, ILA, *Report of the Fifty-Second Conference* (Helsinki 1966), arts IX–XI.

has revolved around environmental matters.[2] Such is the acknowledged importance of such protection that all key instruments of modern international water law contain discrete provisions detailing the precise nature of environmental and ecological obligations applying and the modalities of related inter-State cooperation.[3]

These obligations are today exemplified by Part IV of the 1997 UN Watercourses Convention, which includes express obligations for watercourse States concerning protection of international watercourse ecosystems, prevention, reduction and control of pollution, introduction of alien species, and protection of the marine environment.[4] Indeed, even where instruments predate this general concern with environmental and ecological protection and have not included such provisions, they may either be amended to incorporate such values[5] or interpreted by international courts and tribunals as requiring such protection, especially in the light of the subsequent environmental commitments of the States concerned.[6] In fact, effective implementation of the extensive environmental and ecological protection obligations inherent in general international water law is greatly aided by the parallel elaboration of corresponding rules and principles in international environmental law,

---

[2] *Case Concerning the Gabčíkovo-Nagymaros Project (Hungary/Slovakia)* (1997) *ICJ Reports* 7; *Pulp Mills on the River Uruguay (Argentina v Uruguay)* (Judgment) [2010] *ICJ Reports* 14; *Indus Waters Kishenganga Arbitration (Pakistan v India)* [2013] (Partial Award, 18 February 2013); *Certain Activities Carried Out by Nicaragua in the Border Area (Costa Rica v Nicaragua)* and *Construction of a Road Along the San Juan River (Nicaragua v Costa Rica)* (Judgment on 16 December 2015) (2015) *ICJ Reports* 665.

[3] Notably including the 1997 UN Convention on the Law of the Non-Navigational Uses of International Watercourses (adopted 21 May 1997, entered into force 17 August 2014), (1997) 36 *International Legal Materials* 700; the 1992 UNECE Convention on the Protection of Transboundary Watercourses and International Lakes (adopted 17 March 1992, entered into force 6 October 1996) 1936 UNTS 269; International Law Commission, Draft Articles on Transboundary Aquifers, *Report of the International Law Commission on the Work of Its Sixtieth Session*, (2008) II *Yearbook of the International Law Commission* UN Doc A/CN.4/SER.A/2008/Add.1.

[4] UN Watercourses Convention, arts 20–23.

[5] See, for example, Minute 319 (20 November 2012), amending Treaty Respecting Utilisation of Waters of the Colorado and Tijuana Rivers and of the Rio Grande (3 February 1944) 3 UNTS 314.

[6] *Kishenganga Arbitration* case, *supra*, n. 2.

and their significant normative specificity and procedural sophistication.[7] However, although well established, the environmental requirements of international water law continue to evolve. In recent years, there has been a marked shift in emphasis from the prevention of pollution of shared waters towards the protection of riverine ecosystems, reflecting growing awareness of watercourses as complex and fragile ecosystems providing a range of indispensable ecosystem services and requiring holistic management of a wide variety of interconnected ecological elements. This has obvious implications for the normative scope of international water law, as the so-called 'ecosystem approach' directly links the use of international watercourses to such issues as land use, soil degradation and the protection of aquatic biodiversity. In addition, the utilisation and protection of international watercourses can increasingly affect and be affected by the growing climate change and global biodiversity crises. Indeed, the continuing elaboration of the ecosystem approach and of its constitutive elements may eventually prove crucial to the effective realisation of the fundamental objective of international water law, i.e., the optimal and sustainable use of shared water resources, at a time when the looming problem of freshwater scarcity is increasingly widely recognised as a global environmental crisis.

6.2 Environmental protection in international water law

6.2.1 Pollution Control

The International Law Commission's 1994 Draft Articles,[8] on which the UN Watercourses Convention is based, afforded a very prominent position to environmental obligations. As evidence of the long-standing concern of States with the problem of pollution of international watercourses, the Commission's commentary to the obligation set out in Draft Article 21(2) to prevent, reduce and control pollution of an international watercourse refers to 'a detailed survey of representative illustrations of

---

[7] O. McIntyre, *Environmental Protection of International Watercourses under International Law* (Ashgate, Farnham, 2007), at 379–380.

[8] International Law Commission, *Report of the International Law Commission on the Work of its Forty-Sixth Session*, UN GAOR, Forty-Ninth Sess., Supp. No 10, (1994) UN Doc A/49/10.

international agreements, the work of international organizations, decisions of international courts and tribunals, and other instances of State practice supporting Article 21'. Pollution control requirements have long been centrally relevant to all key aspects of international water law.[9] The Commission commentary regards the due diligence obligation contained in Article 21(2) as 'a specific [pollution related] application of the general principles contained in Articles 5 and 7', respectively addressing equitable and reasonable utilisation and prevention of significant harm. It is immediately obvious that it is 'applying the general obligation of Article 7 to the case of pollution'. However, it is also directly relevant to application of the cardinal principle of equitable and reasonable utilisation as set out under Articles 5 and 6 of the UN Watercourses Convention. For example, Article 5(1) requires that watercourse States utilise shared waters in a sustainable manner 'consistent with adequate protection of the watercourse', which is understood to cover pollution control and the maintenance of riverine ecosystems. Meanwhile, the factors listed in the Convention as relevant to equitable and reasonable utilisation expressly include 'the effects of the use or uses of the watercourses in one watercourse State on other watercourse States'.[10] Similarly, the International Law Commission commentary regards the obligation of watercourse States under Article 21(2) to cooperate in order to harmonise their pollution policies, as being subject to the principle of equitable participation set out under Article 5(2), as well as that of good faith cooperation under Article 8 of the Convention. For the purposes of the pollution control obligations under the UN Watercourses Convention, Article 21(1) defines 'pollution of an international watercourse' very broadly to include 'any detrimental alteration in the composition or quality of the waters of an international watercourse which results directly or indirectly from human conduct'. Thus, Article 21(2) may apply to uses that decrease the flow of a watercourse resulting, for instance, in damage to ecosystems or to salinisation of waters downstream, as well as to more traditional cases of pollution, provided such harm impacts upon another watercourse State. Of course, the obligation enshrined in Articles 21(2) can also be traced to the broader customary obligation for States to prevent transboundary environmental

---

[9] See, for example, *Lac Lanoux Arbitration (Spain v France)* (Award on 16 November 1957) (1961) 24 *International Law Reports* 101.
[10] UN Watercourses Convention, art 6(1)(d). See also art 6(1)(f), referring to 'conservation, protection, development and economy of use of the water resources of the watercourse and the cost of measures taken to that effect'.

harm. Commentators have acknowledged the well-established customary obligation to prevent or abate substantial damage from transfrontier pollution, or any significant risk of causing such damage. The International Law Commission commentary confirms that 'the principle of precautionary action is applicable to the obligation to prevent pollution under Article 21(2), especially in respect of dangerous substances such as those that are toxic, persistent or bio-accumulative'. Similarly, application of Article 21(2) is likely to be informed by the standards adopted under specific watercourse agreements[11] or by other instruments in the field of general international environmental law.[12]

6.2.2 Ecosystems Protection

Whereas pollution control obligations have long been a central element of international water law, the express inclusion in Part IV of the UN Watercourses Convention of an unequivocal obligation regarding the protection of watercourse ecosystems[13] reflects more recent, yet growing awareness among the international community of watercourses as complex and fragile ecosystems, providing a range of indispensable ecosystem services and requiring holistic consideration and management of each watercourse's myriad interconnected ecological elements. The 1994 International Law Commission commentary to Article 20 defines an 'ecosystem' in a manner consistent with prevailing legal and scientific thinking as an 'ecological unit consisting of living and non-living components that are interdependent and function as a community'. Somewhat questionably, the Commission employed the term 'ecosystems of international watercourses' in draft Article 20 with the specific intention of limiting that article's scope of application to the narrow confines of the watercourse itself. The Commission understood the term 'ecosystem' as having 'a more precise scientific and legal meaning' than 'environment' of a watercourse, which it feared 'could be interpreted quite broadly, to apply

---

[11] Including, for example, the 1976 Convention on the Protection of the Rhine against Chemical Pollution, (1977) 16 *International Legal Materials* 242; the 1978 Agreement between the United States and Canada on Great Lakes Water Quality (1978–79) 30 UST 1383.

[12] Including, for example, the 1992 (OSPAR) Convention for the Protection of the Marine Environment of the North-East Atlantic, (1993) 32 *International Legal Materials* 1072; 2001 (Stockholm) Convention on Persistent Organic Pollutants, (2001) 40 *International Legal Materials* 532.

[13] UN Watercourses Convention, art 20.

to areas "surrounding" the watercourses ... [and/or] ... to areas outside the watercourse, which is of course not the intention of the Commission'. However, noting that 'this narrow conception of ecosystem protection is not found elsewhere', leading commentators to doubt whether 'the Commission's careful choice of terminology really does confine the potential scope of this obligation in a meaningful way'.[14] Citing the examples of grazing and logging practices that could have a significant impact on a watercourse, McCaffrey concludes that a meaningful and effective understanding of an 'ecosystem' would almost certainly include 'not only the flora and fauna in and immediately adjacent to a watercourse, but also the natural features within its catchment that have an influence on, or whose degradation could influence, the watercourse'.[15]

Regarding the legal status of ecosystems obligations, the International Law Commission commentary appears to recognise an autonomous customary obligation to preserve and protect watercourse ecosystems, listing a wide range of relevant authorities and declaring that 'there is ample precedent for the obligation contained in Article 20 in the practice of States and the work of international organizations'. However, it also directly links the obligation to protect the ecosystems of international watercourses with the overarching principle of equitable and reasonable utilisation set out in Article 5 of the UN Watercourses Convention by explaining that it 'is a specific application of the requirement contained in Article 5 that watercourse States are to use and develop an international watercourse in a manner that is consistent with adequate protection thereof'. The International Court of Justice similarly considered that Czecho/Slovakia had deprived 'Hungary of its right to an equitable and reasonable share of the natural resources of the Danube—with the continuing effects of the diversion of these waters on the ecology of the riparian area of the Szigetköz'.[16] This differs somewhat from the International Law Commission's understanding of the closely connected obligation in Article 21(2) on control of pollution, which is linked to the application of both Articles 5 and 7, and of Article 22 on the introduction of alien species, which the Commission advises 'should be kept in harmony with

---

[14] P. Birnie, A.E. Boyle and C. Redgwell, *International Law and the Environment* (3rd ed.) (Oxford University Press, Oxford, 2009), at 558–559.
[15] S.C. McCaffrey, *The Law of International Watercourses: Non-Navigational Uses* (Oxford University Press, Oxford, 2001), at 393.
[16] *Gabčíkovo-Nagymaros* case, *supra*, n. 2, at para 85.

the general rule contained in Article 7'. This view may reflect the fact that, unlike Article 20, Articles 21 and 22 appear more concerned with activities potentially causing direct significant environmental harm to other watercourse States, and less concerned with the allocation of quantum share of water.

6.2.3 Ecosystem Approach

Increased ecological awareness has led to 'the adoption of less economic-oriented criteria for the management of freshwater resources, following an "ecosystem approach"',[17] which 'requires consideration of the whole system rather than individual components'.[18] For some time prior to conclusion of the UN Watercourses Convention, a trend towards broader ecosystems obligations was evident in global elaboration of the environmental aspects of international water law. This is apparent from an examination of both basin agreements[19] and regional framework conventions[20] concerning shared transboundary water resources. The 1992 UNECE Water Convention, the only other globally applicable framework convention relating to shared international freshwater resources, provides a notable example of a (formerly) regional instrument, setting out extensive and detailed provisions for the conservation and restoration of the ecosystems of shared basins.[21] Early guidelines adopted under the UNECE Water Convention elaborate upon the meaning and implications of the so-called 'ecosystem approach'.[22] As a framework convention

---

[17] A. Tanzi and M. Arcari, *The United Nations Convention on the Law of International Watercourses* (Kluwer Law International, Dordrecht, 2001), at 8–9.

[18] J. Brunnée and S.J. Toope, 'Environmental Security and Freshwater Resources: A Case for International Ecosystem Law', (1994) 5 *Yearbook of International Environmental Law* 41, at 55.

[19] See, for example, Great Lakes Water Quality Agreement between Canada and the United States (adopted and entered into force 22 November 1978) 30 UST 1383, TIAS No 9257, arts I and II; the Agreement on Cooperation for Sustainable Development of the Mekong River Basin, (1995) 34 *International Legal Materials* 864, arts 3 and 7.

[20] See, for example, the original Protocol on Shared Watercourse Systems in the Southern African Development Community (SADC) Region, arts 2(3), 2(11) and 2(12), (28 August 1995).

[21] *Supra*, n. 3, arts 1(2), 2(2)(b), 2(2)(d) and 3(1)(i).

[22] UNECE, *Guidelines on the Ecosystem Approach in Water Management* (December 1993) UN Doc ECE/ENVWA/31.

originally applying across the wider European region, the ecosystem protection provisions of the UNECE Water Convention have inspired a number of subsequently adopted European river basin agreements, which demonstrate a correspondingly broad commitment to ecosystem protection.[23] The emerging rules on shared international groundwater resources would appear to be evolving in a similar vein with the International Law Commission's 2008 Draft Articles on Transboundary Aquifers stressing 'the role of the aquifer or aquifer system in the related ecosystem' and calling upon states 'to ensure that the quantity and quality of water retained in an aquifer or aquifer system, as well as that discharged through its discharge zones, are sufficient to protect and preserve such ecosystems'.[24]

6.2.4 Environmental Flows

Despite some continuing uncertainty, particular features of an ecosystem approach as applied in the specific context of international water law are beginning to come to light. Most notably, emerging requirements to maintain minimum environmental flows will play a central role in effective implementation of an ecosystem approach in transboundary basins. Environmental flows are intended to provide 'a methodological approach that incorporates environmental concerns into the process of allocating water rights among different users',[25] where the overriding objective 'is to modify the magnitude and timing of flow releases from water infrastructure (e.g., dams) to restore natural or normative flow regimes that benefit

---

[23] Including, inter alia, the Convention on the Protection of the Rhine (22 January 1998), arts 2, 3 and 5; the Convention on Cooperation for the Protection and Sustainable Use of the Danube River (29 June 1994), arts 1(c), 2(30 and 2(5); the Agreements on the Protection of the Meuse and Scheldt (26 April 1994), art 3; and the Framework Agreement on the Sava River Basin (3 December 2002) 2367 UNTS 688, art 11(a).

[24] International Law Commission, Draft Articles on Transboundary Aquifers, *Report of the International Law Commission on the Work of Its Sixtieth Session*, (2008) II *Yearbook of the International Law Commission* UN Doc A/CN.4/SER.A/2008/Add.1, arts 5(1)(i) and 10. See also UNECE, Model Rules on Transboundary Groundwaters (2014), Provision 2.1.

[25] S. Brels, D. Coates and F. Loures, *Transboundary Water Resources Management: The Role of International Watercourse Agreements in Implementation of the CBD* (CBD Secretariat, 2008), at 13.

downstream river reaches and their riparian ecosystems'.[26] The concept is defined in a soft-law instrument as 'the quantity, timing and quality of water flows required to sustain freshwater and estuarine ecosystems and the human livelihoods and well-being that depend on these ecosystems'.[27] Although the issue of environmental flows is seldom addressed directly in international water instruments, its legal character must be understood as 'part of a broader notion of taking an ecosystem approach', and so 'the relevant international instruments are not only those directly dealing with water resources, but also those that have a primary focus on the protection of nature and ecosystems'.[28] A requirement to maintain minimum environmental flows, derived from more established normative principles of international environmental law, received ground-breaking judicial support in the *Kishenganga Arbitration* before a Permanent Court of Arbitration Tribunal, which concluded that 'hydro-electric projects ... must be planned, built and operated with environmental sustainability [and minimum environmental flow in particular] in mind' on several possible legal grounds.[29] At an earlier stage in the proceedings, the Tribunal granted Pakistan's request for interim measures, thereby preventing India from conducting any 'permanent work on or above the riverbed that may inhibit the restoration of the full flow of that river to its natural channel'. In the *Certain Activities* case the International Court of Justice similarly recognised the legal significance of maintaining flow for ecological purposes.[30] A 2011 study assessing State and treaty practice notes that 'the need to provide environmental flows in order to conserve ecological integrity of water basins is becoming more and more important',[31] while a 2013 analysis conducted by a broad coalition of international actors concludes that 'there is now wide recognition of the

---

[26] N. LeRoy Poff and J.H. Matthews, 'Environmental Flows in the Anthropocene: Past Progress and Future Prospects', (2013) 5–6 *Current Opinion in Environmental Sustainability* 1, at 1.

[27] International Water Centre, The Brisbane Declaration (3–6 September 2007).

[28] M. Dyson, G. Bergkamp and J. Scanlon (eds.), *Flow: The Essentials of Environmental Flows* (IUCN, 2003), at 87–88.

[29] *Kishenganga Arbitration* case, (Partial Award), *supra*, n. 2, at paras 450–452 and 454; and (Final Award), (20 December 2013).

[30] *Certain Activities* case, *supra*, n. 2.

[31] G. Aguilar and A. Iza, *Governance of Shared Waters: Legal and Institutional Issues* (IUCN, Gland, 2011), at 99.

importance of maintaining an appropriate flow regime to maintain the ecological health of river basins'.[32]

As the legal nature of the obligation to maintain flows becomes ever clearer, by means of the emerging practice of international courts, water convention secretariats, and national regulatory authorities, the science is similarly advancing for 'the quantification of the linkages between hydrological processes and components and various ecological variables'.[33] Commentators identify certain 'guiding elements' for environmental flows, including the need to describe flow-ecology and flow-social relationships, the need to engage stakeholders in setting environmental water objectives, and the need to integrate environmental flow considerations into infrastructure planning and operation. Beyond water convention regimes, certain multilateral environmental convention secretariats have produced technical guidance on aspects of the calculation and implementation of environmental flow requirements,[34] as have leading environmental civil society organisations. Academic research similarly continues to develop and refine environmental flow methodologies.

6.2.5 Ecosystem Services

The overarching objective of an ecosystem approach, and thus of any regime for maintaining environmental flows, appears increasingly to centre around the concept of ecosystem services, which serves to enhance awareness of the nature and value of socially beneficial services provided by natural ecosystems, and to provide a methodology for valuation and consideration of such services within the decision-making processes of international water law. The 2005 Millennium Ecosystem Assessment provides an essential typology of four categories of ecosystem services, comprising supporting services, provisioning services, regulating ser-

---

[32] R. Speed *et al*, *Basin Water Allocation Planning: Principles, Procedures and Approaches for Basin Allocation Planning* (UNESCO, Paris, 2013), at 58.
[33] See, for example, N. LeRoy Poff, R.E. Tharma and A.H. Arthington, 'Evolution of Environmental Flows Assessment Science, Principles, and Methodologies', in A.C. Horne *et al* (eds.), *Water for the Environment: From Policy and Science to Implementation and Management* (Elsevier, 2017), 203–236, at 203.
[34] See, for example, J. Adams, *Determination and Implementations of Environmental Water Requirements for Estuaries* (Ramsar Convention Secretariat/Secretariat of the CBD, 2012).

vices and cultural services,[35] which can assist in transboundary water cooperation by providing watercourse States with a common understanding of the costs and benefits for each State of measures for the utilisation and protection of shared watercourse ecosystems. In this way, the ecosystem services concept improves the prospects for the agreement of benefit-sharing arrangements among watercourse States, potentially leading to both optimised utilisation and more effective protection of shared watercourse ecosystems. Similarly, accepted methodologies for valuing ecosystem services may facilitate more focused consideration of potential State responsibility for transboundary ecological harm.

Despite the absence of a formally elaborated legal framework for ecosystem services within international water law, the use of such methodologies is becoming more common in the practice of transboundary water cooperation. The Mekong River Commission, for example, recognising the direct linkage between ecosystem components and services, has developed an approach to ecosystem management that 'can involve an assessment of the ecosystem components and/or an assessment of the ecosystem services that are derived from the interaction of those components in support of human well-being'.[36] Guidance on water resources management for the maintenance of ecosystem services has also been developed under the auspices of the Ramsar Convention and the Convention on Biological Diversity (CBD). The CBD Strategic Plan for Biodiversity 2011–20 confirmed that a key purpose of biodiversity conservation is that of safeguarding ecosystem services essential for human well-being and includes a number of targets relating to the ecosystem services provided by inland waters.[37] It appears that, despite few examples thus far of the adoption of formal legal arrangements between States regarding ecosystem services, State practice regarding shared watercourses clearly recognises the linkage between upstream stewardship of ecosystem services and downstream beneficial utilisation of such services. In addition, an emerging legal obligation to maintain ecosystem

---

[35] Millennium Ecosystem Assessment, *Ecosystems and Human Wellbeing: Synthesis* (Island Press, New York, 2005), at 39–48.

[36] Mekong River Commission, *The ISH Study 0306 Study: Development of Guidelines for Hydropower Impact Mitigation and Risk Management in the Lower Mekong Mainstream and Tributaries* (MRC, December 2015), vol 2, at 214.

[37] Convention on Biological Diversity, *Strategic Plan for Biodiversity 2011–2020* (CBD Decision X/2, 2010), Annex, para 13.

services is supported by recent statements framing the issue in the context of a human right of access to ecosystem services.[38]

6.2.6 Adaptive Management

As regards certain important mechanisms for implementing an ecosystem approach, such as that of broad stakeholder participation, general international water law appears as yet rather underdeveloped, although transboundary water institutions tend, in practice, to demonstrate a degree of flexibility in allowing for public participation. This is particularly true, however, of the paradigm of adaptive management, which is widely regarded as central to effective application of an ecosystem approach, and involves a strategy that is 'iterative and flexible, responsive to the constantly changing conditions of both complex ecosystem processes and available scientific knowledge'.[39] Adaptive management is necessary to cope with uncertainty regarding the functions of complex dynamic socio-ecological systems, the value of certain ecosystems and their services, and the potential effects of certain policies and projects on the functioning of ecosystems. We might expect such uncertainty to be significantly exacerbated, and adaptive strategies to become ever more necessary, in the light of the threat represented to freshwater ecosystems by increased climate variability. Stated simply, adaptive management seeks to ensure the resilience of an ecosystem, i.e., the ability of that system to cope with inevitable changes, by adopting a systematic approach for adapting and improving natural resources management by learning from previous management interventions. However, beyond the intrinsic flexibility of the normatively indeterminate principle of equitable and reasonable utilisation, incorporation of adaptive measures into conventional systems of legal rules is problematic, largely due to traditional prioritisation of the stability of legal regimes over their flexibility, especially where such regimes are intended to facilitate investment in large-scale water infrastructure. Thus, traditional legal frameworks for natural resources management tend to be linear, whereas the complexity and dynamism

---

[38] *Report of the Special Rapporteur on the Issue of Human Rights Obligations Relating to the Enjoyment of a Safe, Clean, Healthy and Sustainable Environment* (19 January 2017) UN Doc A/HRC34/49, at 4.
[39] V. De Lucia, 'Competing Narratives and Complex Genealogies: The Ecosystem Approach in International Environmental Law', (2015) 27 *Journal of Environmental Law* 91–117, at 93.

of interconnected ecosystems and social systems require flexibility and a measure of discretion, which is not often afforded to cooperative transboundary institutions (where they exist) by sovereign watercourse States. States inevitably tend to be reluctant to surrender sovereign control over water resources to joint institutions charged with implementing adaptive management. Exceptions to the rule might include the 1996 Farraka Treaty[40] and the 2002 Incomati-Maputo Agreement,[41] as agreements that create mechanisms for the mutually agreed adjustment of flows during times of drought and flooding.

Therefore, legal frameworks for transboundary cooperation must evolve to create suitably empowered and capacitated institutions employing highly sophisticated procedures for inter-State engagement over shared water resources. Legal arrangements reflecting such an approach would accommodate uncertainty through flexible decision-making procedures which permit 'incremental and gradual changes that transition experimentally to new standards or arrangements, while monitoring, assessing and adjusting these changes and their effects'.[42] Although this will present significant challenges for the procedural and institutional arrangements currently prevailing in international water law, the requirement for adaptive resilience governance is not without some legal authority. Strong links exist between adaptive management and the precautionary principle, as both seek to accommodate scientific uncertainty, and the former can be regarded as a means of implementing the latter, which enjoys extensive support as customary law.[43] The precautionary principle is also commonly understood to be an integral element informing application of the ecosystem approach, which can itself be legally justified as a precautionary measure. Of course, the ecosystem approach may already

---

[40] Treaty on Sharing of the Ganga Waters at Farakka (New Dehli, 12 December 1996), arts II and X.

[41] Tripartite Interim Agreement for Cooperation on the Protection and Sustainable Utilisation of the Water Resources of the Incomati and Maputo Watercourses (Johannesburg, 29 August 2002), art 10 and Annex I.

[42] E. Raitanen, 'Legal Weaknesses and Windows of Opportunity in Transnational Biodiversity Protection: As Seen Through the Lens of an Ecosystem Approach-Based Paradigm', in S. Maljean-Dubois (ed.), *The Effectiveness of Environmental Law* (Intersentia, Cambridge, 2017) 81–100, at 96.

[43] See O. McIntyre and T. Mosedale, 'The Precautionary Principle as a Norm of Customary International Law', (1997) 9 *Journal of Environmental Law* 221–241.

enjoy autonomous legal authority, at least in the field of international watercourses. Consistent International Court of Justice endorsement in transboundary watercourses cases of a requirement for continuing environmental assessment might amount to judicial recognition of the important role of adaptive ecosystem-based management in certain situations of scientific uncertainty. The Court stated unequivocally in *Pulp Mills* that 'once operations have started and, where necessary, throughout the life of the project, continuous monitoring of its effects on the environment shall be undertaken',[44] thereby building upon Judge Weeramantry's earlier endorsement in *Gabčíkovo-Nagymaros* of the 'principle of continuing environmental impact assessment'.[45]

6.2.7 Stakeholder and Public Participation

Although Rio Principle 10 proclaims a general principle of public participation,[46] equally applicable to transboundary water resources management and arguably reflecting customary international law, few water agreements expressly require the involvement of stakeholders or the wider public. International water law thus appears somewhat out of step with wider developments in general international law. Water agreements typically focus on inter-State engagement to the exclusion of public participation, as epitomised by Part III of the UN Watercourses Convention. Similarly, Article 9 of the Convention only provides for the regular exchange of data and information at the inter-State level, neglecting public or stakeholder access. Although the UNECE Water Convention is considered progressive regarding stakeholder participation, it only requires parties to make relevant information available to the public,[47] saying little about public participation generally. Some basin agreements inspired by the UNECE Water Convention take a similarly restrictive approach as regards public or stakeholder participation,[48] while others are more inclusive.[49] There exist a limited number of basin agreements

---

[44] *Pulp Mills* case, *supra*, n. 2, para 205.
[45] *Gabčíkovo-Nagymaros* case, *supra*, n. 2, Separate Opinion of Judge Weeramantry, at 108–110.
[46] UN Conference on Environment and Development (UNCED), Rio Declaration on Environment and Development, (1992) 31 *International Legal Materials* 874, UN Doc A/CONF.151/26 (vol 1).
[47] UNECE Water Convention, art 16.
[48] E.g. Danube Convention, art 14.
[49] E.g. Rhine Convention, art 14.

from other regions, most notably Africa, that expressly stipulate public consultation.[50] Where permanent institutional structures for transboundary water management are in place, many assist in facilitating structured stakeholder engagement regardless of the formal terms of the institutional charter given its wide recognition as important for effective ecosystems protection.

Despite limited treaty practice providing for public participation, many regimes either require[51] or promote[52] reliance upon environmental impact assessment (EIA) of planned projects to avoid and minimise adverse impacts and facilitate meaningful inter-State notification. The ICJ has described the obligation to conduct EIA of major projects as a 'requirement under general international law',[53] thereby suggesting that its inclusion in conventional instruments isn't required. While the Court found that 'no legal obligation to consult the affected populations arises for the Parties from the instruments invoked',[54] it suggested that States should rely on their domestic EIA laws to give effect to this obligation, which regimes would almost universally include public or stakeholder participation as a central element.

Public participation is recognised as central to the ecosystem approach in practice developed under the Convention on Biological Biodiversity.[55] Of 12 principles identified by CBD COP 5 to guide implementation of the ecosystem approach, Principle 12 recommends the involvement of all sectors of society, while Principle 11 exhorts decision-makers to make use of all forms of information, including indigenous knowledge.[56] If effective

---

[50] Agreement on the Establishment of the Zambezi Water Commission (13 July 2004), art 16(8); Convention on Sustainable Management of Lake Tanganyika ('Lake Tanganyika Convention') (12 June 2003), arts 5(2)(d), 17.
[51] Lake Tanganyika Convention, art 15.
[52] UN Watercourses Convention, art 12; ILC, Draft Articles on Transboundary Aquifers, art 15(2).
[53] *Pulp Mills* case, n. 2, para 204.
[54] *Ibid.*, para 216.
[55] Biodiversity Convention Decision V/6, Principles 11 and 12; CBD's Revised Programme of Work on Inland Water Biological Diversity, CBD Decision VII/4 (2004), Goal 2.5, Annex 22; Biodiversity Convention Decision VII/11 (13 April 2004), UN Doc UNEP/CBD/COP.7/21, para 10.
[56] *Ibid.*; Biodiversity Convention Decision VII/4 (2004), Goal 2.5 of the CBD's Revised Programme of Work on Inland Water Biological Diversity, Biodiversity Convention Decision VII/4 (2004), Annex, at 22;

public or stakeholder participation is crucial for the protection and preservation of watercourse ecosystems, and thus for achieving optimal and sustainable utilisation of international watercourses, then the prevailing rules on procedural engagement, which are almost exclusively focused on inter-State communication, are not fit for purpose. While participatory rights are developing rapidly within the related fields of human rights law and environmental law,[57] implementation of an ecosystem approach will demand progressive advances in the procedural rules employed in international water treaties and institutions, and corresponding development of basin institutions.

6.3 The influence of international environmental law

6.3.1 Multilateral Environmental Agreements

The incorporation of ecosystem protection requirements into international water law reflects the continuing evolution of international environmental law more generally, with the conservation of ecosystems identified as a key objective of the 1992 Convention on Biological Diversity.[58] The fifth meeting of the CBD Conference of the Parties (COP), held in Nairobi in May 2000, endorsed the 'ecosystem approach' as the primary framework for action under the Convention, defining it as 'a strategy for the integrated management of land, water and living resources that promotes conservation and sustainable use in an equitable way'.[59] COP 5 also adopted 12 principles to guide the practical implementation of the ecosystem approach including, for example: that 'conservation of ecosystem structure and functioning ... should be a priority

---

[56] Biodiversity Convention Decision VII/11 (13 April 2004), UN Doc UNEP/CBD/COP.7/21, para 10.
[57] Convention on Accession to Information, Public Participation in Decision-Making and Access to Justice in Environmental Matters ('Aarhus Convention') (25 June 1998, entered into force 30 October 2001) 2161 UNTS 447; Regional Agreement on Access to Information, Public Participation and Justice in Environmental Matters in Latin America and the Caribbean ('Escazú Agreement'), (9 April 2018, entered into force, 22 April 2021).
[58] Convention on Biological Diversity (5 June 1992) 1760 UNTS 79, art 1.
[59] Biodiversity Convention Decision V/6, 'Ecosystem Approach' (Nairobi 22 June 2000), UN Doc UNEP/CBD/COP/5/23.

target of the ecosystem approach' (Principle 5); that 'ecosystems must be managed within the limits of their functioning' (Principle 6); and that 'the ecosystem approach should be undertaken at the appropriate spatial and temporal scales' (Principle 7).[60] While universally adopted MEAs, such as the CBD with 196 parties, play a hugely significant role in the formation of customary rules, the significant interlinkages between the CBD's objectives and protection of transboundary watercourse ecosystems have long been recognised. For example, a 2008 study conducted on behalf of the CBD Secretariat acknowledges that 'the equitable and sustainable allocation and management of water are crucial for maintaining the ecological function of freshwater ecosystems', and details the many synergies between the CBD, on the one hand, and the UN Watercourses Convention and UNECE Water Convention, on the other.[61] However, the same study notes that 'globally, these [freshwater] ecosystems are in serious decline, due largely to the pressures placed upon water by its various users, and the rate of loss of biodiversity in them surpasses that from other major biomes by a considerable margin'. Of course, as a 'framework' convention, the CBD has created a sophisticated regime, including the establishment of specialist institutional structures which play a key role in the ongoing elaboration of the relevant rules. To date, the CBD COP has adopted two successive programmes of work on inland water biodiversity which clarify application of the relevant CBD requirements to watercourse ecosystems.[62] Relevant guidance has been developed under the auspices of the CBD on a range of issues relevant to protection of watercourse ecosystems, such as invasive alien species.[63] International legal measures for the ecological protection of other biomes, such as the marine environment, have been similarly influenced by the ecosystem approach pioneered under the CBD.

---

[60] See also, Biodiversity Convention Decision VII/11 (13 April 2004), UN Doc UNEP/CBD/COP.7/21.

[61] Brels, Coates and Loures, *supra*, n. 25, at 5.

[62] Programme of Work on Biological Diversity of Inland Water Ecosystems, Biodiversity Convention Decision IV/4 (1998), Annex I; Revised Programme of Work on Inland Water Biological Diversity, Biodiversity Convention Decision VII/4 (2004), Annex.

[63] Biodiversity Convention Decision VIII/27 (2006), Alien Species that Threaten Ecosystems, Habitats or Species, art 8(h): Further Consideration of Gaps and Inconsistencies in the International Regulatory Framework.

Equally, lessons learned in the implementation of legal requirements arising under other environmental conventions intended to protect and preserve associated water-related ecosystems can also serve to inform the ecosystems obligations arising in international water law. For example, the wealth of detailed technical guidance developed over the years under the auspices of the 1971 Ramsar Convention[64] can play a significant role in informing the ecosystem obligations articulated under the UN Watercourses Convention. Normative development of the core requirement of 'wise use' of wetlands under the Ramsar regime places great emphasis on the maintenance of the ecological character of wetlands 'through the implementation of ecosystem approaches, within the context of sustainable development'.[65] There exists obvious complementarity between these two instruments,[66] with wetlands playing a critical role in the functioning of aquatic ecosystems and in the provision of important ecosystem services, such as flood control, wildlife habitat, groundwater recharge and the 'protection, purification, retention and provision of water resources for water and food supplies ... on which the well-being of people and their livelihoods depend'.[67] Clearly international water law has much to learn from a related conventional regime that has had over 50 years to develop and mature normatively.

6.3.2 Equity and Proportionality

The International Law Commission commentary suggests that Article 20 of the UN Watercourses Convention, on ecosystems protection, is subject to the principle of proportionality, which is commonly regarded as an

---

[64] Convention on the Protection of Wetlands of International Importance, Especially as Waterfowl Habitat ('Ramsar Convention') (opened for signature 2 February 1971, entered into force 21 December 1975) 996 UNTS 245.
[65] Ramsar Conference of the Parties Resolution IX.1 (Kampala 2005), Annex A, para 22.
[66] O. McIntyre, 'The Ramsar Convention on Wetlands and General International Water Law: Mutually Supportive Regimes', in R.C. Gardiner, R. Caddell and E. Okuno (eds), *Wetlands and International Environmental Law: The Evolution and Impact of the Ramsar Convention* (Edward Elgar, Cheltenham, 2023).
[67] Ramsar Conference of the Parties Resolution IX.3 on Engagement of the Ramsar Convention on Wetlands in Ongoing Multilateral Processes Dealing with Water (Kampala, 2005), at 1, para 3.

element or device of equity[68] or, more specifically, as 'one technique ... to achieve an equitable outcome in the face of special geographic circumstances'.[69] The commentary to Article 21 takes a similar approach, reflecting the overarching and pervasive significance of the requirement for equitable balancing of interests under the Convention. The Commission's commentary also links ecosystems protection under Article 20 to the principle of sustainable development, which is in turn closely associated with equitable and reasonable utilisation. In what might amount to an early water-related legal articulation of the ecosystem services concept, the commentary explains that 'protection and preservation of aquatic ecosystems help to ensure their continued viability as life support systems, thus providing an essential basis for sustainable development'. Such thinking helps to explain the general significance of a broad, distributive conception of equity throughout the entire normative framework provided by international water law.

6.3.3 Precaution

Consistent with established international practice in relation to ecosystems obligations, the International Law Commission commentary recognises a clear link between ecosystems protection and the precautionary principle, stating categorically that 'the obligation to protect the ecosystems of international watercourses is thus a general application of the principle of precautionary action'. In *Gabčíkovo-Nagymaros*, the ICJ also appears implicitly to have linked the precautionary principle with the ecosystem approach as applied to international watercourses.[70] In addition, by including a specific obligation to protect watercourse ecosystems, regardless of any transboundary impact, Article 20 can be regarded as being inherently precautionary and capable of facilitating action to prevent harmful activity before it is too late to restore ecological integrity. Having regard to the complex science and the resulting level of scientific uncertainty surrounding the processes involved in aquatic ecosystem functioning, damage and remediation, it seems reasonable to assume

---

[68] International Court of Justice, *North Sea Continental Shelf* cases *(Federal Republic of Germany v Denmark and Federal Republic of Germany v Netherlands)* (1969) *ICJ Reports* 3, at 52.

[69] R. Higgins, *Problems and Process: International Law and How We Use It* (Clarendon, Oxford, 1994), at 230.

[70] *Gabčíkovo-Nagymaros* case, *supra*, n. 2, at para 140.

that Article 20 requires that watercourse States take preventive measures against threats of serious or irreversible harm to watercourse ecosystems, even in the absence of conclusive scientific evidence of the likelihood or inevitability of such harm.

6.3.4 Due Diligence

As with all substantive obligations of international water law, Article 20 of the UN Watercourses Convention creates an obligation of 'due diligence', i.e., one relating to the reasonable conduct to be expected of States, rather than an absolute obligation as to result. Therefore, it requires States whose activities may potentially impact upon the ecosystem of an international watercourse to take all appropriate measures to protect and preserve the ecosystem of that watercourse. The standard of conduct expected of States will be informed by 'any standards and practices applicable in the region, among the States in question, or among States of a comparable level of development', and may require, in the light of the continuing evolution of the precautionary principle, 'the establishment of holistic programs of watercourse protection, which should be proactive and anticipatory rather than reactive and remedial in nature'.[71] Such programmes might include measures for 'the protection of land areas associated with watercourses, protection of biological diversity in the riverine environment, and the maintenance of minimum stream flows'. The relevant examples of treaty and soft law instruments cited in the International Law Commission commentary suggest that Article 20 creates obligations regarding polluting activities and the management of fisheries impacting on ecological properties of shared waters, as well as more far-reaching obligations regarding land use practices impacting upon a watercourse. However, Article 20 is likely to have its greatest impact regarding water quantum and flow rather than water quality, with an obligation emerging for watercourse States to ensure the minimum flow required in order to maintain riverine ecosystems.[72] Although neither Article 20 nor the Commission's commentary thereto makes any explicit mention of the impact of alterations to the quantity or flow of water in a watercourse, it has long been apparent that the consequences of damming or diverting an international watercourse might have serious implications for the obligation to protect and preserve the ecosystem of international watercourses. It is clear that

---

[71] McCaffrey, *supra*, n. 15, at 460–461.
[72] *Kishenganga Arbitration* case, *supra*, n. 2; *Certain Activities* case, *supra*, n. 2.

the due diligence requirements flowing from the duties to control transboundary water pollution and protect watercourse ecosystems comprise both substantive and procedural elements. The substantive requirements include a watercourse State's duty to regulate effectively polluting activities or water abstraction where these might cause harm to other watercourse States or to the watercourse. In *Pulp Mills*, the International Court of Justice stressed the obligation of States to adopt and ensure effective compliance with domestic rules and measures that correspond with applicable international agreements or customary rules.[73] The most significant procedural due diligence requirements include early inter-State notification and consultation and, where necessary, negotiation, in respect of potentially harmful planned projects or uses of international watercourses. Each can only meaningfully be performed in conjunction with an EIA which includes assessment of the transboundary effects of the project in question. Beyond planned measures, a further procedural due diligence requirement concerns the duty of watercourse States to engage in regular exchange of data and information on the condition of an international watercourse. Of course, this might imply related obligations regarding the monitoring of riverine uses or conditions and the compilation and analysis of the results derived therefrom. In the *Pulp Mills* case the Court reiterated the central role of cooperative institutional mechanisms in discharging procedural due diligence requirements and stressed that these obligations must be discharged in 'good faith'.[74]

6.3.5 Invasive Species

Article 20 of the UN Watercourses Convention is supplemented by Article 22, which requires watercourse States to take all necessary measures regarding the introduction of alien or new species which may have effects detrimental to the ecosystem of an international watercourse, and may be regarded as a more specific application of the general ecosystems obligation set out in Article 20 and of the related environmental 'no-harm' obligation set out in Articles 7 and 21. The International Law Commission commentary to Article 22 reflects concern that the introduction of such 'alien or new species of flora or fauna into a watercourse can upset its ecological balance', and explains, somewhat surprisingly, that a separate article is necessary because Article 21 'does not include

---

[73] *Pulp Mills* case, *supra*, n. 2, at paras 195–197.
[74] *Ibid.*, at paras 87–91 and 145–149.

biological alterations'. Article 22 applies to non-native, genetically altered or biologically engineered species of plants, animals or other organisms, including parasites and disease vectors, and again imposes a due diligence obligation on watercourse States. As might be expected in relation to the ecological risks presented by alien or invasive species, the commentary suggests that Article 22 requires 'precautionary action'. Although relatively few watercourse agreements include a similar provision,[75] controls on the introduction of new species are not uncommon in other international environmental instruments.[76] The International Law Commission's 1994 conception of the 'ecological balance' that Article 22 is intended to safeguard, appears to correspond with the more modern focus on ecosystem services, as it warns that introduction of alien species can 'result in serious problems including the clogging of intakes and machinery, the spoiling of recreation, the disruption of food webs, the elimination of other, often valuable species, and the transmission of disease'.

6.3.6 Marine Environment

Article 23 of the UN Watercourses Convention requires watercourse States to take measures with respect to an international watercourse necessary to protect and preserve the marine environment. This amounts to recognition of the increasingly serious problem of pollution that is transported into the marine environment by major watercourses, many of which are international, and reaffirmation of obligations regarding so-called 'land-based sources' of marine pollution long recognised in international marine environmental law,[77] including a set of rules codified in 1972 by the International Law Association.[78] As with the rest of Part IV of the UN Watercourses Convention, Article 23 creates a due diligence obligation for watercourse States, requiring them individually or, where appropriate jointly, 'to take all of the necessary measures of which they are capable, financially and technologically'. Once again, such action is to

---

[75] Article 4(2) of the 2000 Revised SADC Protocol on Shared Watercourse provides one notable exception.
[76] Most notably, art 196 of the 1982 UN Convention on the Law of the Sea (10 December 1982) 1833 UNTS 3, on which art 22 is based.
[77] Most notably, arts 194(3)(a) and 207 of the 1984 UN Convention on the Law of the Sea (UNCLOS), *supra*, n. 76 and Convention for the Prevention of Marine Pollution from Land-Based Sources (4 June 1974) 1546 UNTS 119.
[78] International Law Association, Articles on Marine Pollution of Continental Origin, *Report of the Fifty-Fifth Conference* (New York 1972).

be taken 'on an equitable basis' and, given the interlinkage with Articles 20–22, one may assume that the precautionary principle applies. It is worth noting that recent research and related media publicity concerning the nature and scale of the problem of marine pollution by plastics, and the realisation that as much as 90 per cent of the global input of plastics waste into the marine environment is due to ten major river systems, six of which are transboundary (the Ganges, Indus, Mekong, Amur, Nile and Niger), has shone a spotlight on the contribution of watercourses to marine pollution and on the difficult challenge of crafting and implementing an effective regime in international law for the management of marine pollution from land-based sources.[79]

6.4 Conclusion

It is clear, therefore, that requirements of environmental protection figure prominently within the overall governance framework provided by international water law. This is not alone due to the fact that utilisation of water resources is often dependent upon minimum standards of water quality, but also to the highly elaborated content of many environmental rules and principles. The ecosystem approach, as one key normative environmental paradigm, plays an increasingly central role, not only by requiring the maintenance of riverine ecosystems and the beneficial services provided thereby, but also by providing a structured framework that assists in the identification and valuation of the water-related benefits in which watercourse States are to share equitably. In this way, the ecosystem services concept, which is so central to the ecosystem approach, does much to assist reconciliation of the overarching objective of equitable and reasonable utilisation with the imperative of environmental protection— in pursuit of the elusive goal of 'optimal and sustainable utilization' as envisaged under the UN Watercourses Convention.[80]

---

[79] C. Schmidt, T. Krauth and S. Wagner, 'Export of Plastic Debris by Rivers into the Sea', (2017) 51(12) *Environmental Science and Technology* 6634–647. See O. McIntyre, 'International Water Law's Role in Addressing the Problem of Marine Plastic Pollution: A Vital Piece in a Complex Puzzle!', (2022) 6 *Chinese Journal of Environmental Law* 218–252.

[80] UN Watercourses Convention, art 5(1).

7. Institutional arrangements for transboundary water resources management

7.1 Introduction

While international law relating to the management of international watercourses has undergone significant development and clarification in recent years, the institutional machinery required to achieve the requisite level of inter-State cooperation has been developing apace. One key development concerns increasingly widespread adoption of the 'common management' approach, long promoted by learned associations and diplomatic conferences, whereby the basin is regarded as an integrated whole and is managed, to a greater or lesser extent, as an ecological and economic unit, with legitimate interests in the shared waters vested in the community of co-basin States. Such a 'community of interest' approach requires the establishment of institutional machinery to formulate and implement common policies for the management and development of the shared basin. This approach has become all the more necessary due to the complexity of modern water resources utilisation and environmental protection obligations. Such common management institutions, often possessing considerable technical, legal, political and administrative expertise, may be charged with a wide variety of functions, including environmental responsibilities, which are routinely expressly set out in their founding instruments.

Common management is an approach to managing water resources and related problems, rather than a normative principle of international law, and has been widely endorsed by the international community. Recommendation 51 of the *Action Plan for the Human Environment* adopted at the 1972 Stockholm Conference called for the 'creation of river basin commissions or other appropriate machinery for co-operation between interested States for water resources common to more than one jurisdiction' and set down a number of basic principles by which such commissions should be guided.[1] Chapter 18 of Agenda 21, adopted in Rio at the 1992 UN Conference on Environment and Development (UNCED), calls for 'integrated water resources planning and development' and goes on to suggest the role of any institutional machinery established for such purpose.[2] There are estimated to be well over one hundred international river and water resources commissions established to date, and this number is certain to grow in the coming years due to the commitment to institutionalised transboundary water cooperation set out under SDG target 6.5 and measured under SDG indicator 6.5.2.[3]

7.2 Community of interest

The idea that a community of interest exists in an international watercourse, and the related idea that the interests of riparian States can be identified and safeguarded on the basis of equity and solidarity, has long been supported in international judicial deliberation. In a 1929 case concerned with rights of navigation, the Permanent Court of International Justice referred to 'principles governing international fluvial law in general' to conclude that 'this community of interest in a navigable river becomes the basis of a common legal right, the essential features of which are the perfect equality of all riparian States in the use of the whole course of the river and the exclusion of any preferential privilege of any one riparian State in relation to the others'.[4] In the same passage, the Court refers to 'the possibility of fulfilling the requirements of justice and the consider-

---

[1] 1972 Stockholm Action Plan for the Human Environment, UNCHE, *Report of the UN Conference on the Human Environment* (5–16 June 1972) UN Doc A/CONF.48/14/Rev.1, Rec. 51, ch II, s B. See also, *Report of the UN Water Conference, Mar del Plata, 14–25 March 1997* (1977).

[2] UNCED, *Report of the United Nations Conference on Environment and Development, Rio de Janeiro, 3–14 June 1992* (1992) UN Doc A/CONF.151/26 (vol 2), 167, paras 18.3 and 169, para 18.10.

[3] See O. McIntyre, 'SDG 6: Ensure Availability and Sustainable Management of Water and Sanitation for All', in I. Bantekas and F. Seatzu (eds.), *Commentary on the Sustainable Development Goals* (Oxford University Press, Oxford, 2023) 441–508, at 482–486.

[4] *Case Relating to the Territorial Jurisdiction of the International Commission of the River Oder* (Judgment No 16) (1929) *PCIJ Reports*, Series A No 23, 5–46, at 27–28.

ations of utility', anticipating a role for considerations of equity in giving effective protection to the rights of States. In the *Gabčíkovo-Nagymaros* case, the International Court of Justice (ICJ) reproduced this passage from the *River Oder* judgment and stated that 'modern development of international law has strengthened this principle for non-navigational uses of international watercourses as well'.[5] On this basis, a leading commentator suggests that 'the concept of community of interest can function not only as a theoretical basis of the law of international watercourses, but also as a principle that informs concrete obligations of riparian States, such as that of equitable utilization'.[6] Where a community of interest approach is adopted and implemented by means of common management institutions, 'a State's "interests" in an international watercourse system would generally be defined by its present and prospective uses of the watercourse as well as its concern for the health of the watercourse ecosystem'.[7]

The concept of community of interest is traced back to a 1792 French decree on the opening of the Scheldt River to navigation, a position quickly adopted in a number of instruments concerned primarily with rights of navigation in international rivers. Recognising the interdependence of the Rhine States, the 1815 Vienna Congress led to the establishment of the Central Commission for Navigation on the Rhine, which was not only the first international river basin organisation, but also the first ever international organisation. Early agreements articulating the community of interest concept were not restricted to navigational uses of watercourses. Article 4 of the 1905 Treaty of Karlstad between Sweden and Norway provides that international lakes and watercourses 'shall be considered as common'. In modern treaty practice, Article 1(2) of the original 1995 Southern African Development Community (SADC) Protocol on Shared Watercourse Systems provided that the Member States are to 'respect and abide by the principles of community of interests in the equitable utilization of those systems and related resources',[8] though the superseding 2000 Revised SADC Protocol on Shared

---

[5] *Case Concerning the Gabčíkovo-Nagymaros Project (Hungary/Slovakia)* (1997) *ICJ Reports* 7, para 85.
[6] S.C. McCaffrey, *The Law of International Watercourses: Non-Navigational Uses* (Oxford University Press, Oxford, 2001), at 152.
[7] *Ibid.*, at 165.
[8] Food and Agriculture Organisation, *Treaties Concerning the Non-Navigational Uses of International Watercourses: Africa* (FAO Legislative Study 61, 1997) 146.

Watercourses,[9] contains no corresponding provision, but rather follows the approach taken under the 1997 UN Watercourses Convention.[10] Nevertheless, Southern Africa has witnessed renewed efforts to establish basin-wide cooperative institutions. For example, Article 1(2) of the 1992 Agreement between Namibia and South Africa on the Establishment of a Permanent Water Commission provides that the Commission's objective is, inter alia, 'to act as technical adviser to the Parties on matters relating to the development and utilization of *water resources of common interest to the Parties*'.[11] It is more usual, however, for modern treaties 'to *treat* international watercourses as being of common interest than to *refer* to them expressly as common rivers or property'.[12]

Expert commentators have long advocated the community of interest principle and use of the associated common management approach, though few would contend that such an approach has evolved, or is likely soon to evolve, into a requirement of customary international law. It follows, therefore, that despite the extent of inter-State cooperation regarding international watercourses, reliance upon common management through joint institutions can only be established by treaty, rather than under the rules of customary international law.

7.3 Common management institutions

Common management regimes must, therefore, be voluntary arrangements, established by treaty between basin States. General international law will not impose a positive obligation compelling basin States to create such institutions for joint utilisation. The commentary to Article 64 of the International Law Association's (ILA) 2004 Berlin Rules on

---

[9] See the revised Protocol on Shared Watercourse Systems in the Southern African Development Community (SADC) Region, (2001) 40 *International Legal Materials* 321.

[10] United Nations Convention on the Non-Navigational Uses of International Watercourses, (1997) 36 *International Legal Materials* 700 ('UN Watercourses Convention').

[11] See (OSPAR) Convention for the Protection of the Marine Environment of the North-East Atlantic, (1993) 32 *International Legal Materials* 1147 (emphasis added).

[12] McCaffrey, *supra*, n. 6, at 158.

Water Resources, which requires, '[w]hen necessary', the establishment of 'a basin-wide or joint agency or commission with authority to undertake the integrated management of waters of an international drainage basin', nevertheless concedes that 'customary international law does not specifically require [that] such institutions be established nor does it provide specific details for such mechanisms'.[13] Studies of practice relating to international basin management organisations and their founding agreements identify a wide range of issue areas with which such organisations might be concerned, and their organisational structure will vary greatly depending, inter alia, on the range of issue areas covered, the powers and mandate of the institution and the degree of integration and cooperation envisaged by the riparian States. Nevertheless, all international water management institutions appear, formally or effectively, to employ decision-making mechanisms requiring unanimous vote or consensus.

Though common management arrangements must be entered into by States voluntarily, the accumulated practice of States in participating in such arrangements might serve to bolster the normative status, in customary or general international law, of the various rules of procedural engagement comprising the general duty to cooperate, ubiquitous in international water agreements and endorsed by Article 8 of the UN Watercourses Convention. This obligation is widely understood as consisting of a number of specific procedural obligations, such as the duty to notify, the duty to consult and/or negotiate in good faith, and the ongoing exchange of water-related information. State practice in relation to common management might inform the normative content of such rules by demonstrating that bona fide participation in common management institutions contributes to satisfaction of the obligations inherent therein.

The 1992 UNECE Water Convention unusually requires parties to 'enter into bilateral or multilateral agreements or other arrangements' which 'shall provide for the establishment of joint bodies' having a wide range of environmental tasks.[14] Furthermore, common management

---

[13] International Law Association, 'Berlin Rules on Water Resources Law', in *Report of the Seventy-First Conference of the International Law Association* (ILA, Berlin, 2004).

[14] 1992 UNECE Convention on the Protection and Use of Transboundary Watercourses and International Lakes, (1992) 31 *International Legal Materials* 1312, art 9(1) and (2).

must become an ever greater imperative as recognition of the physical unity of the drainage basin gains ground in international law through the ongoing elaboration of the 'ecosystem approach' to the management of international watercourses. The 1997 UN Watercourses Convention expressly encourages watercourse States to enter into common management arrangements. The principle of 'equitable participation' set out under Article 5(2) suggests the nature and scope of the role potentially to be played by joint mechanisms. The ILC commentary explains that Article 5(2) involves 'not only the right to utilize an international watercourse, but also the duty to cooperate actively with other watercourse States in the protection and development of the watercourse',[15] which might, in certain circumstances, require quite intense co-operation. Therefore, a riparian State wishing to claim its right to equitable and reasonable utilisation should consider carefully any invitation to participate in a regional water body or river basin commission. Regarding the general obligation of watercourse States under Article 8 to cooperate 'in order to attain optimal utilization and adequate protection of an international watercourse', Article 8(2) expressly proposes the use of joint mechanisms and commissions. The explicit reference to 'the establishment of joint mechanisms or commissions' in Article 8(2) was not included in the 1994 ILC Draft Articles, but inserted later, perhaps signalling growing acceptance of the common management approach and growing awareness of its merits. Such arrangements would also facilitate the regular exchange of data and information required under Article 9 of the UN Watercourses Convention. Considering the kinds of information listed under Article 9(1), it is apparent that regular and effective exchange of such information, facilitated by common management institutions, would play a significant role in determining an equitable regime for the use or development of an international watercourse. This supports the principle of equitable and reasonable utilisation as elaborated under Articles 5 and 6 of the Convention and helps to ensure that environmental issues are properly considered.

Common management institutions have an obvious role regarding the measures expected of watercourse States under Article 21 of the UN Watercourses Convention in relation to the 'prevention, reduction and

---

[15] International Law Commission, *Report of the International Law Commission on the Work of its Forty-Sixth Session*, UN GAOR, Forty-Ninth Sess., Supp. No 10, (1994) UN Doc A/49/10, at 220.

control of pollution' including, inter alia, 'setting joint water quality objectives and criteria'. Furthermore, Article 24, on the 'management' of international watercourses, provides that 'watercourse States shall, at the request of any of them, enter into consultations concerning the management of an international watercourse, which may include the establishment of a joint management mechanism', thereby suggesting yet again the efficacy of using permanent common management institutions, especially for the purpose of developing the watercourse in an environmentally sustainable manner. Once again, however, the 1994 ILC commentary to Article 24 emphasises that it 'does not require ... that they establish a joint organization, such as a commission, or other management mechanism'.[16] Finally, the UN Watercourses Convention envisages a role for common management mechanisms in the settlement of inter-State disputes, providing that parties may 'make use, as appropriate, of any joint watercourse institution that may have been established by them'.[17]

At a more general level, the notion that all riparian States have a community of interest in an international watercourse reinforces the limited conception of territorial sovereignty upon which the principle of equitable and reasonable utilisation rests and 'expresses more accurately the normative consequences of the physical fact that a watercourse is, after all, a unity' and thus 'implies collective, or joint action' and 'evokes shared governance'.[18] Indeed, it might be argued that, in the absence of common management arrangements, the core traditional substantive rules of international watercourses law may be of limited avail in handling complex problems of water scarcity and quality. Measures intended to realise equitable and reasonable utilisation of shared water resources, an inherently legally indeterminate principle, can be greatly assisted by expert inter-State institutional machinery. Discussing such 'sophist principles' of international law, Franck observes that they 'usually require an effective, credible, institutionalized, and legitimate interpreter of the rule's meaning in various instances'.[19] The related international water law objectives of optimal and sustainable use of shared waters, which

---

[16]   Ibid., at 301.
[17]   UN Watercourses Convention, art 33(1).
[18]   McCaffrey, *supra*, n. 6, at 168
[19]   T.M. Franck, *Fairness in International Law and Institutions* (Oxford, Clarendon, 1995), at 67 and 81–82.

have risen to greater prominence in recent years,[20] equally depend upon active participation and intense cooperation among riparian States in the joint and integrated management of the shared watercourse. Clearly, such procedural engagement amongst States can only be facilitated by means of established permanent institutional machinery, possessing the scientific and technical expertise allowing them to provide impartial specialist advice. The effectiveness of common management machinery for the purpose of effective cooperative management of international watercourses has long been apparent, with such institutions among the earliest and longest established. Early examples include the Commission of the River Rhine established at the 1815 Congress of Vienna, (but only made operational by the 1868 Treaty of Mannheim), the Danube Commission established in 1878, and the International Boundary and Water Commission established by the US and Mexico in 1889.

7.4 International river commissions and environmental protection

Though common management institutions can vary greatly in terms of their composition and function, almost all possess the requisite technical skills, resources and mandate to pursue environmental protection. The 1994 Meuse and Scheldt Agreements, for example, each create an international commission to facilitate cooperation between the parties for the purpose of the environmental protection of the two rivers.[21] Similarly, the 1994 Danube Convention establishes an international commission to facilitate cooperation in order to 'at least maintain and improve the current environmental and water quality conditions of the Danube River and of the waters in its catchment area and to prevent and reduce as far as possible adverse impacts and changes occurring or likely to be caused'.[22] The Danube Commission has more specific functions including, where appropriate, the establishment of emission limits applicable to individ-

---

[20] UN Watercourses Convention, art 5(1).
[21] 1994 Agreements on the Protection of the Rivers Meuse and Scheldt, (1995) 34 *International Legal Materials* 851 and 859, art 2(2).
[22] 1994 Convention on Cooperation for the Protection and Sustainable Use of the Danube River, (1994) 5 *Yearbook of International Environmental Law* 16, arts 2(2) and 4.

ual industrial sectors, the prevention of the release of hazardous substances and the definition of water quality objectives.[23] The US-Canada International Joint Commission (IJC) is one of the longest established such bodies and has developed a comprehensive body of environmental practice regarding the use of shared freshwaters. The IJC was established by the 1909 Boundary Waters Treaty to issue approvals regarding the use, obstruction, or diversion of shared boundary waters and to investigate specific issues where requested. The IJC has issued numerous landmark reports on environmental matters.[24]

The need for joint institutions is particularly acute in relation to environmental protection of watercourses, where the increasingly common practice of establishing international joint commissions creates technically competent inter-governmental bodies to identify and address environmental concerns, along with formal procedural mechanisms for their communication. Such bodies inevitably bring environmental considerations to the fore. However, the advent of the so-called 'ecosystem approach', with its far-reaching implications for international watercourse utilisation and cooperation, makes such bodies ever more necessary.

7.4.1 Ecosystem Approach

In recent years, many international instruments concerning international watercourses have sought to move beyond the traditional obligations to utilise an international watercourse in an equitable and reasonable manner and to prevent significant transboundary harm, and increasingly focus upon 'purely' environmental obligations, including the adoption of a more ecosystem-oriented approach to such protection. This tendency has been greatly advanced by the 1997 UN Watercourses Convention. Reflecting emerging scientific understanding, it expressly requires parties to the Convention to act to protect and preserve international watercourse ecosystems.[25] Though it is beyond the scope of this chapter to elaborate fully upon the 'elusive, unstable and contested' normative implications

---

[23]   Ibid., art 7.
[24]   IJC, *Transboundary Implications of the Garrison Diversion Unit* (1977); IJC, *Water Quality in the Poplar River Basin* (1981); IJC, *Impacts of a Proposed Coal Mine in the Flathead River Basin* (1988).
[25]   UN Watercourses Convention, art 20.

of the ecosystem approach[26] as applied in international water law, some consensus is emerging regarding certain of its core elements, and these create such challenges for inter-State procedural cooperation that permanent and highly capacitated institutional frameworks become all the more necessary. In order to understand better the nature of the procedural challenges, it is necessary to examine several of these elements in turn.

7.4.1.1 *Environmental flows*

Central to an ecosystem approach to the protection of an international watercourse is the establishment of a regime of 'environmental flows', which 'is increasingly accepted as essential for addressing issues of river health, sustainable development, and the sharing of benefits between users'[27] The environmental flow concept is defined by the International Union for Conservation of Nature (IUCN) as 'the water regime provided within a river, wetland or coastal zone to maintain ecosystems and their benefits where there are competing uses and where flows are regulated', thereby providing 'a flow regime that is adequate in terms of quantity, quality and timing for sustaining the health of the rivers and river systems', having regard to relevant social and economic factors.[28] Unfortunately, the issue of environmental flows is seldom addressed directly in international water instruments, so its legal character must be understood in the context of a broader commitment to the adoption an ecosystem approach to shared water resources management. In determining minimum environmental flows, 'the relevant international instruments are not only those directly dealing with water resources, but also those that have a primary focus on the protection of nature and ecosystems'.[29] A binding requirement to maintain minimum environmental flows, derived from broader established principles of international environmental law, has received ground-breaking judicial support from a Permanent Court of

---

[26] V. De Lucia, 'Competing Narratives and Complex Genealogies: The Ecosystem Approach in International Environmental Law', (2015) 27 *Journal of Environmental Law* 91–117, at 91.

[27] J. Scanlon and A. Iza, 'International Legal Foundations for Environmental Flows', (2003) 14 *Yearbook of International Environmental Law* 81, at 83.

[28] M. Dyson, G. Bergkamp and J. Scanlon, *Flow: The Essentials of Environmental Flows* (IUCN, Gland, 2003), at 3–5.

[29] Scanlon and Iza, *supra*, n. 27, at 87–88.

Arbitration tribunal in the *Kishenganga Arbitration*,[30] and the ICJ has recently recognised the legal significance of maintaining flows for ecological purposes.[31] The emergence of a related legal obligation is supported by a number of in-depth legal studies[32] and others recognising 'the importance of maintaining an appropriate flow regime to maintain the ecological health of river basins'.[33] As the legal nature of this obligation becomes clearer, through the combined practice of international courts, water convention secretariats, and national agencies, the scientific and technical aspects are similarly advancing.[34] A growing body of technical guidance is emerging on implementing environmental flow requirements. In terms of legal clarity, commentators have identified certain 'guiding elements' for environmental flow methodologies.[35] Clearly, the task of identifying minimum environmental flows and of crafting and implementing water management arrangements to maintain such flows will present a major challenge for transboundary water cooperation and will require highly capacitated and robust institutional mechanisms.

7.4.1.2 Ecosystem services

As noted elsewhere in this volume, the overarching objective of an ecosystem approach, and thus of any regime for maintaining environmental flows, centres around the ecosystem services concept, which aims to promote understanding of the nature and value of socially beneficial services provided by natural ecosystems, and to provide a methodology for

---

[30] Permanent Court of Arbitration (PCA), *Indus Waters Kishenganga Arb-itration (Pakistan v India)* [2013] (Partial Award, 18 February 2013), at paras 450–452 and 454, and (Final Award), 20 December 2013.

[31] *Certain Activities Carried Out by Nicaragua in the Border Area (Costa Rica v Nicaragua)* (2015) *ICJ Reports* 665, at para 105, Request for Provisional Measures Order, 8 March 2011, Separate Opinion of Judge Sepulveda-Amor, at para 25; Declaration of Judge Greenwood, at para 15.

[32] G. Aguilar Rochas and A. Iza, *Governance of Shared Waters: Legal and Institutional Issues* (IUCN, Gland, 2011), at 99.

[33] R. Speed, et al, *Basin Water Allocation Planning: Principles, Procedures and Approaches for Basin Allocation Planning* (UNESCO, Paris, 2013), at 58.

[34] N. LeRoy Poff, R.E. Tharma and A.H. Arthington, 'Evolution of Environmental Flows Assessment Science, Principles, and Methodologies', in A.C. Horne, et al (eds.), *Water for the Environment: From Policy and Science to Implementation and Management* (Academic Press, Elsevier, 2017) 203–236, at 203.

[35] *Ibid.*, at 225–227.

their valuation and adequate consideration within legal decision-making processes. The 2005 Millennium Ecosystem Assessment provides an essential typology of ecosystem services,[36] which can assist in transboundary water cooperation by providing watercourse States with a common understanding of the costs and benefits for each State of measures for the utilisation and protection of shared watercourse ecosystems. In this way, the ecosystem services concept enhances the prospects for agreement over utilisation and benefit-sharing arrangements amongst watercourse States, potentially leading to both optimised utilisation and more effective protection of shared watercourse ecosystems. Accepted methodologies for valuing ecosystem services might also facilitate consideration of State responsibility for transboundary ecological harm. In either role, the ecosystem approach can function to facilitate the avoidance and resolution of inter-State water resources disputes.

Despite the dearth of attention paid to ecosystem services in international water resources instruments or other formal expressions of the law related to transboundary freshwater ecosystems, such methodologies are increasingly commonly employed in the practice of transboundary water cooperation. Guidance on water resources management for the maintenance of ecosystem services has been developed under the auspices of both the Ramsar Convention[37] and the Convention on Biological Diversity (CBD),[38] and the latter regime confirms the safeguarding of essential ecosystem services as a priority objective in its Strategic Plans.[39] Recent statements of the UN Special Rapporteur on Human Rights and the Environment support an emerging legal obligation to maintain ecosystem services, framing the issue as one involving a human right of access to ecosystem services.[40]

---

[36] Consisting of four categories: supporting, provisioning, regulating and cultural services. See Millennium Ecosystem Assessment, *Ecosystems and Human Wellbeing: Synthesis* (Island Press, New York, 2005), at 39–48.

[37] D. Russi, et al, *The Economics of Ecosystems and Biodiversity (TEEB) for Water and Wetlands* (IEEP/Ramsar Convention Secretariat, 2013).

[38] CBD's Revised Programme of Work on Inland Water Biological Diversity, CBD Decision VII/4 (2004), Annex.

[39] Convention on Biological Diversity, *Strategic Plan for Biodiversity 2011–2020* (CBD Decision X/2, 2010), Annex, at para 13.

[40] *Report of the Special Rapporteur on the Issue of Human Rights Obligations Relating to the Enjoyment of a Safe, Clean, Healthy and Sustainable Environment* (19 January 2017) UN Doc A/HRC34/49, at 4.

Discussion of ecosystem services often includes consideration of the role of payment for ecosystem services (PES) and, though this concept is not extensively developed in international legal practice, key actors in the field of transboundary water management provide some guidance on how such payment systems might work.[41] PES can inform inter-State engagement over transboundary waters because the inter-linkage between the upstream provision of services, often involving improved environmental or water resources stewardship, and the downstream utilisation of services thus provided is increasingly widely recognised and understood to operate on an extensive, transboundary scale. PES arrangements may be utilised as one element of integrated benefit-sharing arrangements and can provide a potentially useful means of rebalancing competing State interests in a shared watercourse. Benefit-sharing will commonly include context-specific arrangements for redistribution of benefits, for compensation for benefits foregone or for the costs of improved stewardship.

Clearly, consideration of ecosystem services, and even arrangements for payment or compensation for the provision of such services, in the determination of an equitable and reasonable regime utilisation regime, which might in turn include complex benefit-sharing trade-offs, will require highly capacitated and robust institutional mechanisms in which co-riparian States could place considerable trust.

7.4.1.3 *Adaptive management*

It is now generally accepted among scholars that iterative, flexible and responsive 'adaptive management' strategies, which seek to ensure the 'resilience' of an ecosystem by adopting a systematic approach for adapting and improving natural resources management by learning from previous management interventions, play a central role in effective application of an ecosystem approach.[42] The uncertainty that adaptive management seeks to address is greatly exacerbated by the threat of climate change. However, incorporation of adaptive measures into conventional frameworks is highly problematic due to a traditional prioritisation of the

---

[41] UNECE, *Guidance on Water and Adaptation to Climate Change* (2009); IUCN, *PAY: Establishing Payments for Watershed Services* (2006).
[42] Biodiversity Convention Decision V/6, 'Ecosystem Approach' (Nairobi 22 June 2000), UN Doc UNEP/CBD/COP/5/23, at 1.

stability of applicable legal rules, especially where they function to facilitate large-scale investment in an international watercourse.

The procedural rules of international water law, especially those concerned with inter-State notification and engagement over 'planned measures', are very well established[43] and tend increasingly to be based upon 'front-loaded' environmental impact assessment (EIA) studies informing 'one-time' project approval processes, as evidenced by the judicial focus upon EIA.[44] Legal arrangements reflecting an adaptive approach, however, would need to accommodate uncertainty through flexible decision-making procedures facilitating 'incremental and gradual changes that transition experimentally to new standards or arrangements, while monitoring, assessing and adjusting these changes and their effects'.[45]

Robust and sophisticated inter-governmental cooperative institutions, enjoying adequate decision-making mandates and technical resources, are needed to implement adaptive management of transboundary basins. Therefore, legal frameworks for transboundary cooperation must evolve to create suitably empowered and capacitated institutions employing highly sophisticated procedures for inter-State engagement over shared water resources.

Though this will present significant challenges for the procedural and institutional arrangements currently prevailing, the requirement for adaptive governance is not without some authority in international water law. Adaptive management is closely linked to the precautionary principle, as both approaches seek to accommodate scientific uncertainty and the former can be a means of implementing the latter, which enjoys

---

[43] Consider the detailed provisions contained in Part III of the UN Watercourses Convention and the detailed procedures adopted under several basin agreements, such as the ZAMCOM Procedures for Notification of Planned Measures (23 February 2017).
[44] *Pulp Mills on the River Uruguay (Argentina v Uruguay)* (Judgment) [2010] ICJ Reports 14, at para 204.
[45] E. Raitanen, 'Legal Weaknesses and Windows of Opportunity in Transnational Biodiversity Protection: As Seen Through the Lens of an Ecosystem Approach-Based Paradigm', in Maljean-Dubois (ed.), *The Effectiveness of International Law* (Intersentia, Cambridge, 2017) 81–100, at 96.

extensive support as customary law.[46] The ecosystem approach may already enjoy autonomous legal authority, at least in the field of international watercourses.[47] Consistent ICJ endorsement of a requirement for 'continuing' environmental assessment in transboundary watercourses cases might amount to tacit judicial recognition of the role of adaptive ecosystem-based management in certain situations of scientific uncertainty.[48]

7.4.1.4 Stakeholder and public participation

Despite the endorsement in Rio Principle 10 of a generally applicable principle of public participation,[49] and consistent academic support for its customary status,[50] few international water management regimes expressly require the participative involvement of stakeholders or the public, suggesting that international water law is out of step with developments in general international law. Water agreements tend to focus exclusively upon inter-State engagement, as illustrated by Part III of the UN Watercourses Convention. Similarly, Article 9 of the Convention only provides for the regular exchange of data and information at the

---

[46] O. McIntyre and T. Mosedale, 'The Precautionary Principle as a Norm of Customary International Law', (1997) 9 *Journal of Environmental Law* 221–241.

[47] A. Trouwborst, 'The Precautionary Principle and the Ecosystem Approach in International Law: Differences, Similarities and Linkages', (2009) 18 *Review of European, Comparative and International Environmental Law* 26–35, at 30.

[48] *Gabčíkovo-Nagymaros Project* case, *supra*, n. 5, Separate Opinion of Judge Weeramantry, at 108–110; *Nuclear Tests* case *(New Zealand v France)* (1995) (Request for an Examination of the Situation in Accordance with Paragraph 63 of the Court, Judgment of 20 December 1974) *ICJ Reports* 457, at 344; Legality of the Use by a State of Nuclear Weapons in Armed Conflict (Advisory Opinion), (1996) *ICJ Reports* 66, at 140; *Pulp Mills* case, *supra*, n. 44, at para 205.

[49] UN Conference on Environment and Development (UNCED), Rio Declaration on Environment and Development, (1992) 31 *International Legal Materials* 874, UN Doc A/CONF.151/26 (vol 1).

[50] J. Razzaque, 'Information, Public Participation and Access to Justice in Environmental Matters', in Shawkat Alam *et al* (eds.), *Routledge Handbook of International Environmental Law* (Routledge, London, 2012), at 140; J. Ebbesson, 'Principle 10: Public Participation', in J.E. Viñuales (ed.), *The Rio Declaration on Environment and Development: A Commentary* (Oxford University Press, Oxford, 2015), at 287.

inter-State level. Though considered highly progressive in this regard, the UNECE Water Convention merely requires State parties to make information relating to the management of transboundary freshwater resources available to the public[51] and says little about public participation. Certain resulting European basin agreements take a similarly restrictive approach as regards public or stakeholder participation,[52] whilst others are more inclusive.[53] A limited number of basin agreements from other regions, most notably in Africa, expressly stipulate a requirement of public consultation.[54] As a matter of practice, however, many existing transboundary water cooperation institutions tend to assist in facilitating stakeholder engagement. River basin organisations (RBOs) in particular tend to engage with a range of key external non-State actors, including civil society, community and research organisations.[55]

In addition, many treaty regimes either require[56] or promote[57] EIA of planned projects in order to avoid and minimise adverse transboundary impacts and to facilitate meaningful inter-State notification, while the ICJ regards the conduct of EIA in respect of a qualifying project as a 'requirement under general international law',[58] thereby suggesting that its stipulation under an applicable treaty isn't required. While the Court hasn't identified a legal requirement to consult the affected populations in the course of an EIA,[59] it suggests that States should rely on their domestic EIA regimes in giving effect to their international obligation to conduct EIA. Of course, such domestic regimes would universally include public

---

[51] *Supra*, n. 14, art 16.
[52] Convention on Cooperation for the Protection and Sustainable Use of the Danube River (Sophia, 29 June 1994), art 14.
[53] Convention on the Protection of the Rhine (Bern, 12 April 1999), art 14.
[54] Agreement on the Establishment of the Zambezi Water Commission (Kasane, 13 July 2004), art 16(8); Convention on Sustainable Management of Lake Tanganyika (Dar es Salaam, 12 June 2003), arts 5(2)(d) and 17.
[55] S. Schmeier, *Governing International Watercourses: River Basin Organisations and the Sustainable Governance of Internationally Shared Rivers and Lakes* (Routledge, London, 2013), at 108.
[56] Lake Tanganyika Convention, *supra*, n. 54, art 15.
[57] UN Watercourses Convention, art 12; International Law Commission, Draft Articles on Transboundary Aquifers, *Report of the International Law Commission on the Work of Its Sixtieth Session* (2008) II *Yearbook of the International Law Commission* UN Doc A/CN.4/SER.A/2008/Add.1.
[58] *Pulp Mills* case, *supra*, n. 44, at para 204.
[59] *Ibid.*, at para 216.

or stakeholder participation as a central element. Public participation is clearly recognised as central to the ecosystem approach in practice developed under the CBD.

If effective public or stakeholder participation is crucial for protection and preservation of watercourse ecosystems, and thus for achieving optimal and sustainable utilisation of international watercourses, then the prevailing paradigm of procedural engagement in international water law, with its almost exclusive focus on inter-State communication, is outdated and no longer fit for purpose. While participatory rights are developing rapidly within the related fields of human rights law and environmental law,[60] it is clear that implementation of the ecosystem approach will demand significant progressive advances in the procedural rules employed in international water law, and corresponding development of the supporting institutional structures required to facilitate effective transboundary water cooperation.

7.4.1.5 *Benefit-sharing arrangements*

As suggested earlier, ever greater focus on maintaining and maximising ecosystems services, and related obligations regarding environmental flows, raises the prospect of extensive reliance upon so-called 'benefit-sharing' arrangements to optimise efficiency in the beneficial use of ever-scarcer water resources whilst maintaining ecosystem integrity. Such arrangements might typically involve some form of payments for benefits, or for costs associated with enhanced stewardship. The ecosystem services concept provides a methodology for economic and social valuation of the benefits of watercourse ecosystems, including non-marketable benefits, and may thereby permit their integration into benefit-sharing arrangements. The 'payment for ecosystem services' paradigm may play a key facilitating role where the benefit to the optimised, and in lieu of which compensation would be owed to another, is that of ecological integrity and/or the ecosystem services accruing from

---

[60] UNECE, Arhus Convention on Accession to Information, Public Participation in Decision-Making and Access to Justice in Environmental Matters (Aarhus, 25 June 1998), (1999) 38 *International Legal Materials* 517; Regional Agreement on Access to Information, Public Participation and Justice in Environmental Matters in Latin America and the Caribbean (Escazú, 4 March 2018).

a functioning transboundary riverine ecosystem. Such methodologies for identification and valuation of benefits provided by international watercourse ecosystems, where widely accepted, can assist transboundary water cooperation by providing 'a common point of departure' for negotiations over possible benefit-sharing.

However, one should not understate the difficulty of the task of crafting complex benefit-sharing arrangements, as well as that of ensuring their effective implementation and management over time This will require a permanent, adequately capacitated and sophisticated regime capable of facilitating intense technical inter-State engagement, which is beyond the capacity of currently established procedural rules and institutional structures. Experience suggests that a highly sophisticated legal and institutional framework for cooperation would be required for formulating feasible proposals. Case-studies of benefit-sharing arrangements emphasise the critical importance of structured engagement with all stakeholders and of the requisite 'institutionalisation' of transboundary water cooperation. Therefore, benefit-sharing arrangements, and particularly those focused on preservation of watercourse ecosystems and maintenance of ecosystem services, will require capacitated, permanent institutions capable of facilitating intense procedural engagement between watercourse States. The implications are clear in terms of the institutional resources likely to be required.

7.5 Conclusion

Although States cannot be compelled to participate in institutional arrangements for transboundary water cooperation, States increasingly do so voluntarily. Such participation assists in establishing compliance with their legal obligation to cooperate in the management of the shared waters. Though institutional mechanisms have many advantages, they can play a particularly significant role in environmental protection of international watercourses. Though such bodies vary greatly in terms of their functions and organisational structure, most enjoy an express environmental mandate under their founding instruments, as well as a technically competent and well-resourced permanent staff. The emergence of the ecosystem approach to the management of international watercourses broadens the range of functions that will inevitably fall to such institutions, thereby increasing States' reliance on joint commissions with trusted technical expertise and established cooperative procedures. As it may involve immensely complex scientific determinations about the likelihood and seriousness of possible ecological impacts, and an expectation that States will protect watercourse ecosystems while seeking equitably to balance ecosystem protection objectives against other factors relevant to equitable and reasonable utilisation, the ecosystem approach necessarily requires robust, intense and adaptable inter-State dialogue. Therefore, the ecosystem approach highlights and emphasises the benefits offered by institutional cooperation, without which it seems unlikely that it could ever be effectively implemented.

8. International water law and legal convergence

Despite the extraordinary proliferation of instruments of international water law since the adoption of the 1966 Helsinki Rules, it appears that a range of forces are acting to maintain the internal coherence of this sub-field of international law, its coherence within the broader field of international environmental and natural resources law, and its position firmly within the international law system. A range of institutions and processes ensure the continuing unitary nature of international water law within a unitary system of international law, notably including the universalist instincts of the International Court of Justice (ICJ), the codification exercises routinely undertaken by the International Law Commission (ILC) and learned associations, as well as the universal, pervasive and indivisible character of increasingly relevant human rights norms. These processes of 'convergence' act to unify and enrich the fabric of the increasingly elaborate and sophisticated complex of rules, principles and institutional structures comprising international water law, while suggesting its growing maturity after nearly 60 years of frenetic evolution and supporting its continuing coherent elaboration.

8.1 Introduction

The broad corpus of rules, principles and objectives comprising international environmental and natural resources law has grown exponentially in the 50 years since the United Nations Conference on the Human Environment (UNCHE)[1] was convened in Stockholm in June 1972, not least due to the adoption of several hundred agreements and other instruments, ranging from highly focused bilateral arrangements through to global framework conventions. These instruments are further supple-

---

[1] 1972 Stockholm Action Plan for the Human Environment, UNCHE, *Report of the UN Conference on the Human Environment* (5–16 June 1972) UN Doc A/CONF.48/14/Rev.1.

mented by the wealth of soft law declarations, resolutions and guidance, best exemplified by the Rio Declaration adopted on the twentieth anniversary of Stockholm. However, serious questions have persisted regarding whether the body of rules establishes a solid basis for cooperation between states, enabling the global community to address our most pressing environmental problems. This is no less true of the sub-field of international water law, where States can choose to rely on uncompromising conceptions of territorial sovereignty to justify a failure to cooperate or to accept responsibility for transboundary harm, and where compliance and enforcement mechanisms are often deficient. In this regard, it might be characterised as 'an immature and underdeveloped body of law'.[2]

To a certain extent international water law, like international environmental law, might be considered a victim of its own success, at least in terms of the treaty process, where there exists a clear risk of treaty proliferation, treaty congestion and legal fragmentation, as a result of which different, and sometimes inconsistent legal instruments and principles might be called upon to address related problems. Such fragmentation can give rise to a range of difficulties beyond normative incoherence and inconsistency. For example, fragmentation within various international environmental and natural resources regimes can lead to overlapping responsibilities and inefficiency in the pursuit of legitimate environmental and social goals, to normative conflicts between actors, and may undermine effective enforcement.[3] Equally, fragmentation between water and environmental regimes and other rule complexes beyond international environmental and natural resources law, such as the human rights, trade and investment regimes, threatens to undermine normative coherence and the effective application of water, environmental and natural resources rules.

Counteracting the threat of fragmentation, however, one can observe the phenomenon of 'convergence' occurring in the ongoing elaboration of international environmental and natural resources law, whereby different

---

[2] F. Francioni, *Realism, Utopia and the Future of International Environmental Law* (EUI Working Papers: Law No 2012/11), at 1.

[3] J. Gupta, C. Vegelin and N. Pouw, 'Lessons Learnt from International Environmental Agreements for the Stockholm+50 Conference: Celebrating 20 Years of INEA', (2022) 22 *International Environmental Agreements* 229–244, at 235–236.

sub-fields appear increasingly to borrow normative forms and approaches from each other, and from beyond international environmental and natural resources law, in an ongoing process of 'interpenetration and cross-fertilisation'.[4] Most commentators welcome this unifying trend, foreseeing positive benefits in terms of environmental and equitable outcomes. Some highlight the potential environmental role of enforcement mechanisms established beyond international environmental and natural resources law, such as the increasing readiness of the World Trade Organization (WTO) Appellate Body or investor-State arbitration tribunals to give due consideration to the legitimate environmental commitments of States, or the growing concern of human rights courts and monitoring bodies with the environmental dimension of the human rights to life, health, and private and family life.[5] Others describe such interactions between different regimes as a process of 'cognitive learning', by means of which 'regimes learn from each other as, for example, with the "CBD-ification" of older conservation agreements; notions developed within the Convention on Biological Diversity (CBD) were then applied within other conservation agreements'.[6] The phenomenon of convergence represents a critical point in the developmental maturity of international water, environmental and natural resources law, as a process of normative consolidation which largely addresses concerns over fragmentation and treaty congestion, and thereby sustains the unitary and systemic character of this field of international law and bolsters its coherent application, often across multiple sectors and media, and helps to reflect the broad scale and interconnectedness of the natural world.

8.2 Convergence in international law

In response to, or perhaps in step with, the phenomenon of 'fragmentation', widely observed in recent decades in connection with the 'diversification and expansion' of international law and the associated

---

[4] A. Cassese, *International Law* (Oxford University Press, Oxford, 2001), at 45.
[5] Francioni, *supra*, n. 2, at 1.
[6] Gupta *et al*, *supra*, n. 3, at 236.

development of increasingly autonomous sub-regimes,[7] it would appear that a counteracting integrative process of 'convergence' is occurring due to the gradual interpenetration and cross-fertilisation of previously distinct sub-fields of international law. In seeking to reconcile the seemingly incompatible phenomena of fragmentation and convergence, one commentator views both as interdependent processes necessitated by an increasingly extensive, sophisticated and specialised legal system, suggesting that, as a system, international law 'must draw upon and recognise differences between regimes, actors, institutions and processes … [but also that] … difference and diversity must be disciplined, regularised, and contained within particular boundaries'.[8] Thus, the process of fragmentation may be viewed as indicative of the success of the system of international law in addressing the problems of increasingly sophisticated inter-State interactions in an increasingly complex world, while convergence serves to restrain international law's disintegration in order to preserve the requisite degree of systemic unity.

Convergence therefore represents the opposite side of the coin to more commonly aired concerns regarding the fragmentation of international law. Whereas fragmentation has been comparatively well studied, both in terms of its origins in the changing structure and processes of international law and its implications for normative consistency and systemic coherence, the various processes of convergence are rather less well understood. However, some of these processes of convergence are readily apparent. Thus, while the phenomenon of fragmentation has been largely associated with the proliferation in recent years of international courts and tribunals, the International Court of Justice may also be recognised as a key agent for maintaining the unity of the international legal order, particularly through a range of procedural means, such as an expansion of its advisory functions. Similarly, the International Law Commission was conceived, ab initio, as an institution to be entrusted with responsibility for 'the progressive development of international law and its *codifica-*

---

[7] International Law Commission, Fragmentation of International Law: Difficulties arising from the Diversification and Expansion of International Law, *Report of the International Law Commission on the Work of its Fifty-Fifth Session*, UN GAOR, Supp. No 3, (2003) UN Doc A/58/10 Chapter X, at 267.

[8] M. Craven, 'Unity Diversity and the Fragmentation of International Law', (2002) XIII *Finnish Yearbook of International* Law 1–31, at 10.

tion,[9] and is specifically tasked with fostering 'more precise formulation and *systematisation* of rules of international law in fields where there already has been extensive State practice, precedent and doctrine'.[10] Other, quasi-constitutional processes of convergence embedded into the fabric of international law include the principle of systemic integration applying to the interpretation of treaties pursuant to Article 33(1)(c) of the Vienna Convention on the Law of Treaties,[11] which has facilitated 'interpenetration and cross-fertilisation', notably between the specialist sub-fields of international water resources law and international environmental law. This principle builds upon other legal maxims traditionally applied by international courts and tribunals to facilitate normatively coherent reasoning, such as *ut res magis valeat quam pereat*, a tenet of treaty interpretation providing that all provisions are to be construed 'in a way that gives meaning to all of them, harmoniously'. It is important to examine the phenomenon of convergence and to seek to highlight the diverse processes which can facilitate unity and coherence amongst water-related sub-fields of international law. Expert commentators have long expressed concern regarding a lack of awareness of parallel legal developments as a leading cause of fragmentation, with the ILC Study Group on Fragmentation noting that:

> Specialized law-making and institution-building tends to take place with relative ignorance of legislative and institutional activities in the adjoining field and of the general principles and practices of international law. The result is conflicts between rules or rule-systems, deviating institutional practices and, possibly the loss of an overall perspective on the law.'[12]

Thus, an awareness of the processes of convergence allows one to consider the potential relevance of rules originating within fields other than international water law and to understand the means by which these might come to apply.

---

[9] United Nations Charter (1945) 1 UNTS xvi, art 13 (emphasis added).
[10] Statute of the International Law Commission (21 November 1947) UNGA Res. 174 (II), art 15 (emphasis added).
[11] Vienna Convention on the Law of Treaties (1969) 1155 UNTS 331.
[12] International Law Association Study Group (2003), *supra*, n. 7, at 11, para 8.

8.3 Forms of convergence

As convergence occurs in step with fragmentation, its characterisation might usefully commence with an examination of the key forms of the latter phenomenon. Some commentators identify two principal varieties: firstly that of normative incompatibility between different sectoral (sub-) fields of international law, such as that routinely emerging between trade and environmental law; and secondly fragmentation of the basic 'systemic rules' that underpin the idea of international law as a unitary system.[13] The first may be described as a matter of 'judicial communication' and thus as a less serious threat to the systemic integrity of international law, as it 'would be open to remedy—by, for example, developing rules of normative hierarchy or institutional machinery for inter-sectoral dialogue'. The second variety of fragmentation, however, threatens the eventual disintegration of international law 'into a series of specialised, project-specific, regimes operating with little or no consistency between them as regards the relevant actors, institutional priorities, modes of settlement or framing suppositions'.[14] Others identify three distinct forms of fragmentation, including: 'substantive fragmentation', which arises where different sub-fields claim a measure of autonomy and risk forming self-contained fragmented regimes; 'institutional proliferation', which has seen a marked increase, not alone in the number of specialist international courts and tribunals, but also in the new enforcement mechanisms established under sectoral treaties to promote compliance with the increasingly specialised regimes created thereunder; and 'methodological fragmentation', where different sub-fields of international law, and the judicial and quasi-judicial bodies established thereunder, develop different methods for the interpretation of treaties and/or the recognition of customary rules.[15] Specific drivers of convergence have been identified which serve to counteract each.

---

[13] Craven, *supra*, n. 8, at 3.
[14] *Ibid.*
[15] M. Andenas, 'Reassertation and Transformation: From Fragmentation to Convergence in International Law', (2015) 46 *Georgetown Journal of International Law* 685, at 694–702.

8.4 Drivers of convergence

The phenomenon of convergence is supported and facilitated by a number of systemically important and centrally located international institutions, by the continuing evolution and expansive application of certain specialist sub-fields of international law, and by established processes employed in the formation of international rules. Most notable amongst the institutional drivers are the universalist instincts of the International Court of Justice in its role as the pre-eminent adjudicatory body concerned with the determination and application of international rules, along with the International Law Commission's tendency towards the identification and promotion of generally accepted normative positions in its central role in the codification and progressive development of international law. Indeed, the general use of codification initiatives as a means of informing and assisting the rational development of critically important and emerging areas of international legal activity, even by learned societies and other bodies enjoying no law-making mandate formally conferred by States, further contributes to incremental progressive convergence around key legal values in a number of fields. Consider, for example, the seminally important role in the development of modern international water resources law of the International Law Association's 1966 Helsinki Rules on the Uses of the Waters of International Rivers.[16] The universal, pervasive and indivisible character of human rights norms ensure that these rules can play a very influential unifying role in the evolution of many sub-fields of international law, not least international environmental and natural resources law.

8.4.1 International Court of Justice

In response to concerns regarding the risk of fragmentation due to the expansion of international law to cover a multitude of new fields, and the related creation of new and highly specialised courts, enforcement mechanisms and other institutional structures, leading authorities have long argued that the ICJ ought to revitalise its role as the principal judicial organ of the United Nations and of the international community more

---

[16] International Law Association, Helsinki Rules on the Uses of the Waters of International Rivers, ILA, *Report of the Fifty-Second Conference* (Helsinki 1966).

generally in order to ensure the unity and coherence of international law, especially as it and other UN organs such as the ILC, are not limited to applying a single treaty regime and so can draw on their wider experience from a range of fields of law and practice. The ICJ, in particular, 'has a long-standing practice and experience ranging across all fields of law and in applying multiple fields of law simultaneously', a position 'strengthened not only by the extensive jurisprudence, clarifying treaty obligations and customary international law, but also by the quality of and respect for that jurisprudence across legal communities'.[17] In further support of the contention that 'the ICJ is uniquely positioned to lead the way in the shift from fragmentation to convergence in international law', Andenas points out that 'the ICJ's authority is particularly strong on general international law, its principles and method' and further notes 'the interaction between the ICJ and the ILC on the formation of customary international law' and the ICJ's influence on the methodological approaches of other UN bodies involved in the elaboration of international rules.[18] In particular, he notes the Court's recent important contributions to customary international law on human rights and environmental law and its recent recognition and development of the doctrines of *erga omnes* and *jus cogens* as illustrative of a transformation in its approach.[19] Such a transformation can only contribute to the phenomenon of convergence with the doctrine of *jus cogens*, for example, widely understood 'as a force binding international subsystems "within a minimal communal sphere"'.[20] In an observation highly pertinent to the continuing evolution of international environmental and natural resources law, Andenas notes that:

> The development of customary international law by the ICJ is now more likely to include human rights law, international trade law, environmental law, and other fields of international law which until recently seemed to fragment into autonomous regimes. The ICJ has provided itself with the tools to contribute to some level of unity and coherence of international law.[21]

This observation is borne out by the recent history of water resources disputes aired before the ICJ, where emerging rules and principles of international environmental law played a significant role in clarify-

---

[17] Andenas, *supra*, n. 15, at 688–689.
[18] Ibid., at 689–690.
[19] Ibid., at 706.
[20] Ibid., at 707.
[21] Ibid., at 732–733.

ing the water-related rights and entitlements of the watercourse States concerned.[22]

8.4.2 International Law Commission

In concert with the role of the Court in 'nudging' international law in the direction of universality,[23] the International Law Commission plays a critical role in the elaboration of multilateral treaties which similarly contribute to legal convergence. This is particularly true of the many instruments which purport to codify and/or progressively develop rules of custom in a particular field, where the Commission's pronouncements 'carry unusual normative force and may help to resolve difficult legal issues'.[24] In addition to the 'presumption of universality' evident in the recognition of customary rules, such a presumption also pervades many of the reports and draft articles produced by the ILC, aimed as they so often are at 'all States' or the 'international community'.[25] Experience appears to suggest that the Commission's 'pre-legislative' texts, which are normally presented to States as a general statement of the applicable law as it has evolved to date, 'are likely to be acceptable to States only if they can be presented in as general terms as possible'.[26] On the other hand, proposed conventional instruments containing significant divergences tend to be less widely ratified. Crawford further highlights the Commission's seminal work on State responsibility, where the proposed rules are elevated 'to a higher level of generality—from 'primary' to 'secondary' rules', a characterisation which contributed to the widespread State acceptance of the resulting 2001 ILC Draft Articles on State Responsibility.[27]

---

[22] *Case Concerning the Gabčíkovo-Nagymaros Project (Hungary/Slovakia)* (1997) ICJ Reports 7; *Pulp Mills on the River Uruguay (Argentina v Uruguay)* (Judgment) [2010] ICJ Reports 14; *Certain Activities Carried Out by Nicaragua in the Border Area (Costa Rica v Nicaragua)* and *Construction of a Road Along the San Juan River (Nicaragua v Costa Rica)* (Judgment on 16 December 2015) (2015) ICJ Reports 665.

[23] J. Crawford, *Chance, Order, Change: The Course of International Law* (Hague Academy of International Law, General Course on Public International Law, 2014), at 341.

[24] Ibid., at 290.

[25] Ibid., at 325.

[26] Ibid., at 340.

[27] International Law Commission, 2001 Draft Articles on the Responsibility of States for Internationally Wrongful Acts, (2001) II *Yearbook of the International Law Commission* (Part Two), (12 December 2001), adopted

8.4.3 International Human Rights Law

The universal character of many human rights values may also contribute to the tendency towards convergence in international law. Crawford notes that reservations to treaties that might in effect 'violate basic guarantees designed to support the attainment of protected rights' might, in the face of objections by other States, be severed from such treaties and rendered inoperable against any objecting State.[28] Such a restriction might also apply to purported reservations to global, regional or basin-level water agreements, relating for example to the equitable apportionment or environmental protection of shared international water resources, where such a reservation might otherwise permit, in certain extreme cases, violation of an immediately effective and non-derogable obligation associated with the human right to water.[29] This may be the type of scenario contemplated by Wilde when extolling the potential role of the ICJ in applying and developing certain human rights, suggesting that the Court 'might "add value" when compared to treatment by a specialist [human rights] tribunal'.[30]

8.5 Convergence and international water law

Recent developments and practice in the field of international water law illustrate the myriad causal factors that appear to be driving convergence in international environmental and natural resources law and assist identification of the diverse legal processes by means of which the forces of unification can occur. Despite the relative dynamism and growing spe-

---

[28] under UNGA Res. 56/83 UN Doc A/56/49 (vol I)/Corr.4. Crawford, *ibid.*, at 341.
Crawford, *supra*, n. 23, at 339.
[29] See, for example, Article para 37(a) of General Comment No 15: The Right to Water (Articles 11 and 12 of the Covenant), adopted by the Committee on Economic, Social and Cultural Rights (CESCR) (20 January 2003) UN Doc E/C.12/2002/11, which confirms the 'core obligation' of States: 'To ensure access to the minimum essential amount of water, that is sufficient and safe for personal and domestic uses to prevent disease'.
[30] R. Wilde, 'Human Rights Beyond Borders at the World Court: The Significance of the International Court of Justice's Jurisprudence on the Extraterritorial Application of International Human Rights Law Treaties', (2013) 12 *Chinese Journal of International Law* 639–677.

cialisation in international water law, it does not appear to be retreating from the generalised parameters of international law into an autonomous and compartmentalised regime. This is all the more remarkable when one considers humanity's unique and total dependence upon water over and above any other natural resource and the particular challenge this presents for inter-State water cooperation. Further, no two international watercourses are remotely similar—hydrologically, environmentally, demographically, historically, economically, socially, developmentally or politically—causing commentators to identify the use of international watercourses as an area of international legal activity particularly 'likely to be influenced by regional postures and implications'.[31]

The dynamism of modern international water law is demonstrated by its rapid evolution, since its first comprehensive codification in 1966,[32] to find expression in three globally relevant instruments[33] along with hundreds of bilateral and basin-level agreements, and its emergence as the most litigated element in recent international environmental and natural resources disputes.[34] In its frenetic development it interacts intensively with other sub-fields of international law. While the distributive equity enshrined at the core of international water law appears increasingly relevant to other normative fields, most notably the inter- and intra-generational equity lying at the heart of the overarching concept of sustainable development, in pronouncing on water resources disputes

---

[31] Crawford, *supra*, n. 23, at 328.
[32] Helsinki Rules (1966), *supra*, n. 16.
[33] Convention on the Protection and Use of Transboundary Rivers and Lakes (adopted 17 March 1992, entered into force 6 October 1996) 1936 UNTS 269 ('UNECE Water Convention'); United Nations Convention on the Non-Navigational Uses of International Watercourses, (1997) 36 *International Legal Materials* 700 ('UN Watercourses Convention'); International Law Commission, Draft Articles on Transboundary Aquifers, *Report of the International Law Commission on the Work of Its Sixtieth Session*, (2008) II *Yearbook of the International Law Commission* UN Doc A/CN.4/SER.A/2008/Add.1.
[34] *Gabčíkovo-Nagymaros Project* case, *supra*, n. 22; *Pulp Mills* case, *supra*, n. 22; *Indus Waters Kishenganga Arbitration (Pakistan v India)* [2013] (Partial Award, 18 February 2013); *Certain Activities* case, *supra*, n. 22; International Court of Justice (ICJ), *Certain Activities Carried Out by Nicaragua in the Border Area (Costa Rica v Nicaragua)*, (Compensation Judgment on 2 February 2018); *Dispute over the Status and Use of the Waters of the Silala (Chile v Bolivia)* (Judgment on 1 December 2022).

international courts and tribunals have enthusiastically borrowed and assimilated values from related complexes of primary substantive rules including, in particular, international biodiversity law and international human rights law. Cassese might have almost had international water law in mind when noting that the 'special bodies of law' that have gradually emerged are at present 'gradually tending to influence one another, or States and international courts are coming to look upon them as part of a whole'.[35] He continues, in a manner reminiscent of the complex interactions evident in the application of the rules of modern international water law:

> Thus, for instance, two bodies of law, namely international rules and guidelines on the protection of the environment and international trade law are increasingly linked to—and, to some extent, made contingent upon the application of—the law of development as well as human rights law.[36]

Of course, in its recent expansive elaboration international water law has also been subjected to the forces of divergence and fragmentation, most notably with the ILC's 2008 adoption of its Draft Articles on the Law of Transboundary Aquifers,[37] which purport to codify the rules applying to shared international groundwater resources, and which many commentators regard as heralding in certain key respects a markedly different approach to the utilisation and environmental protection of transboundary water resources than that set out in the UN Watercourses Convention,[38] which was itself based on earlier meticulous ILC codification of the law, albeit the law applicable to international surface

---

[35] Cassese, *supra*, n. 4, at 45.
[36] Ibid.
[37] Adopted under UNGA Res.63/124, UN Doc A/RES/63/124. See International Law Commission, *Report of the International Law Commission on the Work of Its Sixtieth Session*, UN GAOR, Sixty-Third Sess., Supp. No 10, (2008) UN Doc A/63/10.
[38] See S.C. McCaffrey, 'The International Law Commission Adopts Draft Articles on Transboundary Aquifers', (2009) 103/2 *American Journal of International Law* 272–393; S.C. McCaffrey, 'The International Law Commission's Flawed Draft Articles on the Law of Transboundary Aquifers: the Way Forward', (2011) 36(5) *Water International* 566–572; O. McIntyre, 'International Water Resources Law and the International Law Commission Draft Articles on Transboundary Aquifers: A Missed Opportunity for Cross-Fertilisation?', (2011) 13 *International Community Law Review* 1–18, at 2–3.

waters.[39] Over time, however, the practice of international water law appears to work to promote unity in the management of groundwater and surface water resources. This trend is illustrated by the 2017 decision of the Ministerial Forum of the Parties of the Orange-Senqu River Commission (ORASECOM)—a river basin organisation (RBO) established between Botswana, Lesotho, Namibia and South Africa—to establish a Multi-Country Cooperation Mechanism for the Stampriet Transboundary Aquifer System (STAS MCCM), which lies within the basin. The STAS MCCM will facilitate joint management of the resources of the aquifer and coordination of the requirements applying to surface waters and groundwater resources. For example, ORASECOM will have responsibility for activities related to the STAS, which are now incorporated into ORASECOM's 10-year Integrated Water Resources Management (IWRM) Plan, running from 2015–2024. This is not an isolated example of the incorporation of groundwater within the activities of RBOs. In Southern Africa alone, similar initiatives have recently been adopted in the context of the Limpopo and Zambezi basins. In the context of the second cycle of national reporting on SDG indicator 6.5.2 on 'progress on transboundary water cooperation', the Danube States reported that groundwater bodies of basin-wide importance are incorporated into the river basin management plan of the International Commission for the Protection of the Danube River (ICPDR). Highlighting the potentially significant unifying role of such governance paradigms as the SDGs, the Second Report on SDG indicator 6.5.2 suggests that 'in other instances, it may be necessary to update older [transboundary cooperation] arrangements in order to integrate principles … which account for both surface water and groundwater'.[40]

Through international water law's dynamic and intensive interactions with other sub-fields, it typifies the processes of 'interpenetration' and 'cross-fertilisation' identified by Cassese.[41] Indeed, evolving practice in the field would appear to exhibit each of the means of addressing normative incompatibility suggested by Craven.[42] For example, 'rules of hierarchy',

---

[39] International Law Commission, *Report of the International Law Commission on the Work of its Forty-Sixth Session*, UN GAOR Forty-Ninth Sess., Supp. No 10, (1994) UN Doc A/49/10.
[40] UNECE and UNESCO, *Second Report on SDG Indicator 6.5.2* (2021), at 20.
[41] Cassese, *supra*, n. 4, at 45.
[42] Craven, *supra*, n. 8, at 3.

which are normative in character rather than temporal or conceptual, would appear to be emerging, under which human rights values take priority, followed increasingly closely by ecological values. The human rights values correspond with international water law's traditional emphasis on providing for vital human needs[43] and are increasingly informed normatively by the active global discourse on the human rights of access to adequate water and sanitation,[44] and by the elaborate implementation framework developed for realisation of Sustainable Development Goal (SDG) 6.[45]

The ecological values, in turn, correspond with increasing recognition of the imperative of protecting water-related ecosystems and maintaining the ecosystem services and human benefits provided thereby. Similarly, 'machinery for institutional dialogue' plays an increasingly important role in ensuring such 'permeation' and 'cross-fertilisation'. For example, the Conferences of the Parties and Secretariats of the Convention on Biological Diversity and the Ramsar Convention on Wetlands continue to elaborate detailed guidance on the integration into international water law instruments and practice of the ecosystems-related obligations arising under each of these flagship global environmental conventions.[46]

---

[43] See, for example, the unique priority above all other uses of water accorded water required for 'vital human needs' under Article 10(2) of the UN Watercourses Convention.

[44] Which has evolved rapidly since adoption of UN Committee on Economic, Social and Cultural Rights, General Comment No 15: The Right to Water (arts 11 and 12 of the Covenant) (20 January 2003) UN Doc E/C.12/2002/11.

[45] O. McIntyre, 'SDG 6: Ensure Availability and Sustainable Management of Water and Sanitation for All', in I. Bantekas and F. Seatzu (eds.), *Commentary on the Sustainable Development Goals* (Oxford University Press, Oxford, 2023), at 441–508.

[46] See, for example, Ramsar COP, 'Engagement of the Ramsar Convention on Wetlands in Ongoing Multilateral Processes Dealing with Water' (2005), Resolution IX.3; S. Brels, D. Coates and F. Loures, *Transboundary Water Resources Management: The Role of International Watercourse Agreements in Implementation of the CBD* (CBD Secretariat 2008); J. Adams, *Determination and Implementation of Environmental Water Requirements for Estuaries* (Ramsar Convention Secretariat 2012); D. Russi et al, *The Economics of Ecosystems and Biodiversity (TEEB) for Water and Wetlands* (Institute for European Environmental Policy and Ramsar Convention Secretariat 2013).

Though convergence plays a role in addressing both principal varieties of fragmentation, it appears that 'it is in the potential inconsistencies developing as regards secondary, or "structural" rules, that one finds much of the debate over fragmentation located', including rules relating to 'the source of obligation, the identity of relevant actors, the method by which competing interests are to be weighed, or the basis for responsibility—that seems to call into question the integrity of the whole'.[47] Taking Craven's characterisation of secondary systemic rules as a starting point, the evolution of international water law as a sub-field appears to demonstrate the operation of convergence in respect of just such rules, thereby contributing to the systemic integrity of international law more generally. As regards 'the source of obligation', for example, foundational cases in the area have turned upon determination of the respective limits of States' rights and duties emanating from territorial sovereignty[48] or upon radical approaches to the general rules of treaty interpretation, including the contextually informed principle of 'contemporaneity'[49] and pioneering application of the principle of 'systemic integration'.[50] Others have focused upon the role and character of intergovernmental institutions, in this case a river basin organisation, in the discharge of procedural obligations and elements of due diligence conduct.[51]

As regards 'the identity of relevant actors', international water law struggles within the traditional confines of international law to adapt to emerging customary requirements on facilitating the participation of relevant stakeholders or the wider public,[52] with the UNECE Water Convention, widely considered one of the most progressive instruments in this regard, even though it only formally requires State parties to make information relating to the management of transboundary freshwaters

---

[47] Craven, *supra*, n. 8, at 9.
[48] *Lac Lanoux Arbitration (Spain v France)* (Award on 16 November 1957) (1961) 24 *International Law Reports* 101; *Certain Activities* case (2015), *supra*, n. 34.
[49] *Gabčíkovo-Nagymaros* case, *supra*, n. 22.
[50] *Kishenganga Arbitration* case, *supra*, n. 34.
[51] *Pulp Mills* case, *supra*, n. 22.
[52] UN Conference on Environment and Development (UNCED), Rio Declaration on Environment and Development, (1992) 31 *International Legal Materials* 874, UN Doc A/CONF.151/26 (vol 1), Principle 10; International Law Commission (ILC), Articles on the Prevention of Transboundary Harm from Hazardous Activities, *Report of the International Law Commission*, Fifty-Third Sess., (2001), UN Doc A/56/10, art 13.

available to the public and says little about public participation.[53] More generally, Andenas emphasises the role of the ICJ in expanding the cast of relevant actors, noting that the Court has contributed to 'resolving pressing problems of human rights and environmental law' by 'moving away from the strictly inter-state, non-hierarchical perspective of international law where state consent has put extreme restrictions on jurisdiction, obligations of states and the development of the law'.[54]

Possibly no other sub-field of international law has been so concerned with elaborating 'the method by which competing interests are to be weighed', since such a process comprises the core of equitable and reasonable utilisation, the overarching cardinal principle of international water law. As regards the 'basis for responsibility', an important recent international watercourse case saw the ICJ extend the application of generally applicable secondary rules on State responsibility to a new class of transboundary harm, the loss by one State of the beneficial enjoyment of ecosystem services due to ecological damage caused by another.[55] Thus, despite its specialised and highly context-specific character, international water law appears to counteract threatened fragmentation of the 'systemic rules' of international law through resort to the universally applicable general rules and institutions of international law. Indeed, it might be understood to address the unavoidable complexity of its subject-matter by employing a significant measure of context sensitivity in the application of general rules, typified by the overarching principle of equitable and reasonable utilisation.

Similarly, if one adopts Andenas's characterisation of fragmentation, we can likewise see the forces of convergence at work within the sub-field of international water law in addressing each of his three species. Looking particularly at 'substantive fragmentation' it is notable that many of the key emerging imperatives of international environmental law have become central features of international water law. For example, in the celebrated *Pulp Mills* case the ICJ recognised the customary status and universal applicability of the requirement to conduct environmental impact assessment (EIA) of a major project potentially impacting

---

[53] UNECE Water Convention, *supra*, n. 33, art 16.
[54] Andenas, *supra*, n. 15.
[55] *Certain Activities* case (2015), *supra*, n. 34.

a shared watercourse,[56] and hence its critically important role in discharging the core duty, contained in almost all watercourse agreements (and in environmental conventions more generally), to notify other watercourse States of such planned measures. Equally, obligations relating to the protection of ecosystems and, more specifically, the maintenance of minimum environmental flows[57] and of essential ecosystem services,[58] have become ubiquitous in global and basin-level instruments. Of course, the rather general and vague EIA and ecosystems obligations which have become such centrally important features of international water law are normatively informed by the rich and comprehensive corpus of environmental conventions, declarative instruments and related practice. Commentators note this continuing 'environmentalisation of International Water Law', which reflects the reality of water resources management and 'its multiple interconnections and unitary complexity—of (non-confined) groundwater with surface water, of the watercourses with other biota and ecosystems, with the land mass as well as with marine waters.'[59] Of course, other sub-fields also permeate the normative requirements of international water law including, most notably, the full range of values associated with international human rights law. Regarding the tendency towards the 'humanisation' of international water law, Canelas de Castro notes, in addition to 'the adoption of a human right to water and sanitation, which corresponds to the satisfaction of the most basic human needs', the wide-ranging recognition of extensive procedural rights, including rights of access to information, or participation in decision-making, including in impact assessments and in inter-State water organisations, and of access to judicial or administrative justice.[60]

8.6 Conclusion

While one more often reads of concern regarding the potential fragmentation of international environmental and natural resources law into

---

[56] *Pulp Mills* case, *supra*, n. 22, at para 114.
[57] *Kishenganga Arbitration* case, *supra*, n. 34.
[58] *Certain Activities* case (2015), *supra*, n. 34.
[59] P. Canelas de Castro, 'Trends of Development of International Water Law', (2015) 6 *Beijing Law Review* 285–295, at 286–287.
[60] *Ibid.*, at 289.

increasingly diverse and uncoordinated sub-fields, or regarding its alienation from other, related fields of international law and practice, in reality various processes of convergence tend to act to unify and enrich the fabric of this increasingly extensive, elaborate and sophisticated complex of rules, principles and institutional structures. These processes suggest the growing developmental maturity of modern international environmental and natural resources law, including international water law, after more than 50 years of evolutionary development, whilst also supporting its continuing coherent elaboration. Experience within the specific sub-field of international water law demonstrates how the processes of convergence operate to enrich the available corpus of rules and principles, by making available to decision-makers the normative and methodological resources of related sub-fields and of international law more generally. Where lacunae appear to exist in the substantive and procedural fabric of international water law, practitioners might venture to look beyond the immediately applicable sub-field of international rules for lessons regarding their progressive interpretation and coherent application.

9. International water law and the Sustainable Development Goals—SDG 6

9.1 Introduction

The Sustainable Development Goals (SDGs) seek to build upon the widely acknowledged success of the preceding Millennium Development Goals (MDGs), which operated between 2000 and 2015 and established the first global set of quantifiable targets for key elements of the sustainable development paradigm, which were non-binding in character and relied solely upon the political commitment of States and other actors. The SDGs came into effect on 1 January 2016 with the aim of stimulating collective action for the realisation of sustainable development and of guiding the developmental decisions of State and other actors over the subsequent fifteen years (2016–30). Of the 17 individual goals and 169 targets identified and agreed under an extensive and participative process, SDG 6, which enshrines an unequivocal commitment to 'ensure availability and sustainable management of water and sanitation for all', is without question one of the most significant in terms of its potential impact on environmental and developmental outcomes. It is not alone important as a critical developmental goal in its own right but also due to its extensive impact on almost all other sectoral SDGs including, most notably, SDG 1 (eradication of extreme poverty), SDG 2 (eradication of extreme hunger), SDG 3 (improving health and well-being), SDG 4 (inclusive and quality education), SDG 5 (gender equality), SDG 7 (affordable and clean energy), SDG 11 (sustainable cities and communities), SDG 12 (responsible consumption and production), SDG 13 (climate action), SDG 14 (marine and coastal pollution), and SDG 15 (land-based ecosystems). Indeed, several of these related SDGs make express mention of water in one or more of their associated targets,[1] indicating that sustainable management of water resources is crucial to their achievement. However, SDG 6 is similarly dependent upon certain other SDGs. As is the case with many of the other goals, achievement of SDG 6 will be largely dependent upon measures required to advance the non-sectoral, cross-cutting goal set out in SDG 16 (peace, justice, and strong institutions) and the 'good governance' values and arrangements inherent therein.[2] Similarly, it is clear that advances in responsible consumption and production in line with SDG 12 will play a significant role in achieving sustainable water management.

Although the SDGs, like the MDGs which preceded them, are intended to be non-legally binding, they are likely to transform utterly the dynamic of the processes for formation of national, transnational, and international rules for the realisation of economic, social, and environmental rights and the promotion of human welfare more generally. It is immediately apparent upon an examination of the specific targets associated with SDG 6[3] that efforts to achieve these targets will need to be supported by formal rules of domestic and international water law, environmental law, and human rights law, which will become increasingly sophisticated and prescriptive in terms of their normative implications.[4] At the same time, it is quite clear that SDG 6 and, more particularly, the specific targets and

---

[1] For example, targets listed under SDG 3 aim to 'end ... water-borne diseases', and 'substantially reduce the number of deaths and illnesses from ... water ... pollution and contamination'; SDG 11 target 4 aims to 'significantly reduce the number of deaths and the number of people affected and substantially decrease the direct economic losses ... caused by ... water-related disasters'; SDG 12 target 4 aims to significantly reduce the release of chemicals and wastes to water; and targets listed under SDG 15 seek to ensure conservation, restoration, and sustainable use of inland freshwater ecosystems, including in particular wetlands, in line with obligations under international agreements.

[2] On the intrinsic linkage between such 'good governance' values and realisation of the human right(s) to water and sanitation, which in turn corresponds closely with SDG 6, see O. McIntyre, 'The Human Right to Water as a Creature of Global Administrative Law', (2012) 37(6) *Water International* 654–669.

[3] UN SDG Goals are available at https://sdgs.un.org/goals/goal6.

[4] On the role of international water law, see O. McIntyre, 'International Water Law and Sustainable Development Goal 6: Mutually Reinforcing Paradigms', in D. French and L.J. Kotze (eds.), *Sustainable Development Goals: Law, Theory and Implementation* (Edward Elgar, Cheltenham, 2018), at 173–201.

related indicators set out thereunder, are likely to act as a strong catalyst for the continuing development of international and domestic rules for sustainable water management. The inclusive and participatory process employed for the elaboration and adoption of the SDGs, as well as the institutional and administrative mechanisms emerging for monitoring progress on their implementation, enhance the normative legitimacy of the core values enshrined in socially progressive water-related legal instruments, thereby conferring upon the SDGs the potential to transform the interpretation and continuing development of water law frameworks. For example, the emphasis in SDG target 6.b on ensuring the participation of local communities in water management should do much to promote the, as yet nascent, elaboration of more inclusive and participatory procedural and institutional arrangements for transboundary water cooperation.

SDG 6 can, therefore, impact profoundly on international water law in a range of ways which cohere with existing trends in its continuing evolution. For example, it is likely to impact the continuing discourse in international law on the human right(s) of access to water and sanitation and, thereby, the normative interpretation and application of the related concept of 'vital human needs' commonly employed in international river basin and water resources agreements.[5] SDG 6 also strongly supports a greater focus on ecosystem-based approaches to water management and regulation and promotes participatory water governance at the international level. Significantly, SDG 6 does much to support and encourage greater transboundary water cooperation amongst watercourse States. It is useful, therefore, to highlight the possible legal implications of the SDGs by examining the multi-faceted, mutually supportive interrelationship between individual targets under SDG 6 and international water law.

SDG 6 embodies the solemn commitment of the international community to work towards ensuring, by 2030 at the latest, universal availability of safe and adequate water and sanitation services, along with sustainable management of the water resources on which such services directly

---

[5] Notably, in the United Nations Convention on the Non-Navigational Uses of International Watercourses, (1997) 36 *International Legal Materials* 700 ('UN Watercourses Convention'), art 10(2) and the International Law Commission, Draft Articles on Transboundary Aquifers, *Report of the International Law Commission on the Work of Its Sixtieth Session*, (2008) II *Yearbook of the International Law Commission* UN Doc A/CN.4/SER.A/2008/Add.1, art 5(2).

depend. As such, it is quite clear that international, transnational and domestic legal frameworks will play a central role in shaping the actions necessary for realisation of SDG 6, despite the latter's non-legally binding character. It is equally apparent, however, that the values set out under SDG 6 must inevitably exert significant influence upon the continuing evolution, elaboration. and implementation of legal measures, principles. and approaches related to water services and to the sustainable management of water resources more generally. This is no less true in respect of international legal frameworks applying to shared transboundary water resources, and so it is important to illustrate the close, two-way interrelationship between international law and the SDGs by exploring the mutually supportive approaches embodied in international water, environmental and human rights law and in SDG 6. The former bodies of law continue to develop rules and principles intended to promote a normatively broad and inclusive right to water and to require environmental protection of shared international water resources and the ecosystems dependent thereon. However, the articulation and solemn adoption of SDG 6 by almost the entire international community of States represents a universal formal political commitment to such values, which can only serve further to legitimise and inform such emerging norms.[6] In addition, the development and agreement of a comprehensive set of targets and indicators related to SDG 6 does much to promote a common understanding of water-related entitlements and duties, and to provide clear benchmarks for the realisation of each aspect of the corresponding human right to water.

Of course, one could not reasonably expect the targets adopted under SDG 6 and the indicators developed subsequently to be capable of capturing and fully reflecting the complexity and increasing sophistication of the corresponding legal frameworks. For example, the discourse that has been ongoing now for well over twenty years regarding the emerging right(s) to water and sanitation has identified normative elements, both substantive and procedural,[7] that could not possibly be captured under

---

[6] All 193 UN Member States unanimously adopted the UN General Assembly Resolution containing the SDGs, see, https://sustainabledevelopment.un.org/memberstates.html.

[7] UN Committee on Economic, Social and Cultural Rights, General Comment No 15 (2002), The Right to Water (arts 11 and 12 of the International Covenant on Economic, Social and Cultural Rights) (20 January 2003) UN Doc E/C.12/2002/11, provides a seminal early analysis of

a mere three indicators.[8] Similarly, the emergence of the so-called 'ecosystem approach' in international water law has given rise to a host of increasingly sophisticated technical standards and methodological tools, such as those relating to minimum environmental flows or to ecosystem services and arrangements for payment therefor, which are not even touched upon by indicator 6.6.1. Therefore, it is helpful to map the water governance implications of SDG 6 implementation onto the corresponding international legal framework for water and to speculate upon the likely influence of each upon the other.

The key Declaration contained in the United Nations General Assembly Resolution by means of which the SDGs were adopted[9] very strongly suggests that the SDGs are to be informed and guided by relevant established and emerging norms of international law. In the case of water access and management in line with SDG 6, these norms will principally be found in the fields of international human rights law, environmental law, and water law. The Declaration confirms that 'the new Agenda is guided by the purposes and principles of the Charter of the United Nations, including full respect for international law. It is grounded in the Universal Declaration of Human Rights, international human rights treaties', etc. It goes on to explain that key, seminal soft-law instruments have helped to shape the 2030 Agenda, including the 1992 Rio Declaration on Environment and Development. More generally, the Declaration, as well as the targets enumerated under each SDG, are suffused with language which suggests the central relevance of the values underlying international human rights law, environmental law and natural resources law. For example, the Declaration envisages 'a world of universal respect for human rights and human dignity, the rule of law, justice, equality and non-discrimination' and one 'in which consumption and production patterns and use of all natural resources … are sustainable', and where environmental protection, climate sensitivity and respect for biodiversity are centrally important. Such language immediately calls to mind the existing international legal frameworks which seek to promote and protect such values. These connotations are even more immediate and direct in respect of SDG

---

the normative and governance implications of the purported emergence of the right to water.

[8] See, SDG 6.1.1, SDG 6.2.1a and SDG 6.2.1b.
[9] Transforming Our World: The 2030 Agenda for Sustainable Development (25 September 2015) UNGA Res. 70/1, UN Doc A/RES/70/1.

6, where the Declaration sets out a vision of 'a world where we affirm our commitments regarding the human right(s) to safe drinking water and sanitation and where there is improved hygiene'. This reference to the putative emergence in international human rights law of a human right(s) to water and sanitation, and thus to the associated intense discourse on the legal status and normative implications of such a right(s), is unmistakable.

At the same time, despite its non-binding character, it appears that in adopting the 2030 Agenda the international community was aware of the potential legal significance of this solemn universal statement of formal State support for emerging, and in some cases contested, rules and principles of international law. For example, the Resolution adopting the 2030 Agenda declared on behalf of all UN Member States that 'we reaffirm all the principles of the Rio Declaration on Environment and Development', which includes all of the key principles associated with the field of international environmental law, even though the customary legal status of some of these has not been definitively established.[10] Indeed, where the Declaration states that 'we reaffirm our commitment to international law and emphasize that the Agenda is to be implemented in a manner that is consistent with the rights and obligations of states under international law', it is also careful to 'reaffirm that every State has, and shall freely exercise, full permanent sovereignty over all its wealth, natural resources and economic activity'. The international community appears, therefore, to have been conscious of the possibility that Resolution 70/1 might come to be regarded as sovereign acceptance of certain contested legal paradigms, such as the human right(s) to water and sanitation.[11] This would have clear implications for the 41 States who abstained in the vote on the 2010 UN General Assembly Resolution on the matter of formal

---

[10] Take, for example, the so-called precautionary principle/approach, codified in Rio Principle 15. Arguing against the Principle's customary status, see D. Bodansky, 'Customary (and Not So Customary) International Environmental Law', (1995) 3(1) *Indiana Journal of Global Legal Studies* 105–119; arguing in favour of its customary status, see O. McIntyre and T. Mosedale, 'The Precautionary Principle as a Norm of Customary International Law', (1997) 9 *Journal of Environmental Law* 221–241.

[11] Botswana provides an example of a State that had contested the legally binding nature of a human right to water, before losing a 2011 legal battle over the San and Bakgalagadi peoples' access to water in the Central Kalahari Game Reserve.

international recognition of a human right to water.[12] Even where the 2030 Agenda could not be regarded as tacit State acceptance of emerging international norms for the purposes of identifying applicable rules of customary international law, the targets set out under each SDG, as well as the indicators and the various modalities subsequently developed to assist with implementation, are likely to play a key role in informing the normative content of such emerging norms and in guiding their effective implementation, regardless of their formal source of legal authority. Consider the targets enumerated under SDG 6, which stress, inter alia, the affordability of water services, the protection of vulnerable groups, environmental and ecosystems protection, and public participation rights, thereby highlighting values and objectives which might be expected to guide the elaboration of future international human rights, environmental or water management measures.

Although the SDGs are clearly guided by, and will operate to pursue, the objectives set out under international law frameworks relevant to sustainable development,[13] they appear to embody the commitment of the international community to take action beyond the narrow and often restrictive confines of the international law-making process. Therefore, the SDGs represent a novel approach emerging in international policy-making, which began with the MDGs and proceeds on the basis of political commitment rather than binding legal norms.[14] It appears, therefore, that, by packaging social priorities into an easily understandable set of goals, and by establishing measurable and time-bound objectives, these goals 'help to promote global awareness, political accountability, improved

---

[12] The Human Right to Water and Sanitation (28 July 2010) UNGA Resolution 64/292, UN Doc A/RES/64/292. States abstaining included: Armenia, Australia, Austria, Bosnia and Herzegovina, Botswana, Bulgaria, Canada, Croatia, Cyprus, Czech Republic, Denmark, Estonia, Ethiopia, Greece, Guyana, Iceland, Ireland, Israel, Japan, Kazakhstan, Kenya, Latvia, Lesotho, Lithuania, Luxembourg, Malta, Netherlands, New Zealand, Poland, Republic of Korea, Republic of Moldova, Romania, Slovakia, Sweden, Trinidad and Tobago, Turkey, Ukraine, United Kingdom, United Republic of Tanzania, United States, Zambia.

[13] For example, the Preamble of Resolution 70/1 declares that the SDGs 'seek to realize the human rights of all'.

[14] Aspects of this approach are evident in the adoption of the Paris Agreement on Climate Change (2015) UN Doc FCCC/CP/2015/L.9/Rev.1.

metrics, social feedback, and public pressures'.[15] This permits the international community to identity goals and targets which are universal, yet aspirational in nature, and which facilitate, more easily than formal international legal measures, engagement with a range of actors crucial to developmental outcomes, including civil society, the scientific community and the private sector. While the 17 goals and 169 targets are global in terms of their scope of application, they are also flexible and articulated in such a way as to be capable of taking into account different levels of national capacity and development, while respecting national policies and priorities.[16] However, it would be mistaken to assume that the SDGs are completely divorced from or immune to developments in relevant areas of international law, which are likely both to inform the implementation of measures designed to realise the SDGs and, in turn, to be supported and enhanced by the implementation of such measures.

In order to illustrate this complex two-way relationship between SDG 6 and relevant rules of international law, it is helpful to select several of the most relevant targets in order to examine how each relates to emerging and established normative frameworks of international law applying to access to water-related services and to the management of shared international water resources and related ecosystems.

9.2 SDG target 6.1: 'by 2030, universal access to safe and affordable drinking water'

The objectives set out under target 6.1 very closely reflect those encompassed by the human right to water and originally set out in 2002 by the UN Committee on Economic Social and Cultural Rights (CESCR) on the basis of the Committee's progressive interpretation of the 1966 International Covenant on Economic, Social and Cultural Rights (ICESCR). By emphasising 'universal and equitable access to safe and affordable drinking water' in target 6.1, the 2030 Agenda mirrors the normative content of the right to water identified in General Comment

---

[15] J.D. Sachs, 'From Millennium Development Goals to Sustainable Development Goals', (2012) 379 *The Lancet* 2206.
[16] See UNECE, *A Healthy Link: The Protocol on Water and Health and the Sustainable Development Goals* (2016) UN Doc ECE/INF/NONE/2016/16, at 3.

No 15, which focuses on three key factors: availability, quality, and accessibility.[17] Under General Comment No 15, 'availability' refers to a supply of water that is 'sufficient and continuous for personal and domestic uses', while 'quality' refers to the safety of the water supply in terms of being 'free from micro-organisms, chemical substances and radiological hazards that constitute a threat to a person's health'. The goal of 'accessibility' as outlined in General Comment No 15 corresponds very closely with the requirements of equitable access and affordability contained in target 6.1, as it comprises four overlapping dimensions: physical accessibility, economic accessibility, non-discrimination, and information accessibility. 'Economic accessibility' requires that 'water and water facilities and services must be affordable to all', while the requirement of 'non-discrimination' dictates that 'water and water facilities and services must be accessible to all, including the most vulnerable or marginalized sections of the population, in law and in fact'. The equitability of access will also depend on 'physical accessibility', meaning that 'adequate water facilities and services must be within safe physical reach for all sections of the population' and in particular, 'must be accessible within, or in the immediate vicinity of, each household, educational institution and workplace'. Similarly, equitable access will be highly dependent upon the corresponding 'right to seek, receive and impart information concerning water issues'. Therefore, the broad aims set out in target 6.1 have already been elaborated in detail in an authoritative, if contested, interpretation of the normative content of a key instrument of international human rights law, and in certain instruments in the field of international water and environmental law.[18]

---

[17] See generally, O. McIntyre, 'The Emergence of Standards Regarding the Right of Access to Water and Sanitation', in S. Turner *et al* (eds.), *Environmental Rights: The Development of Standards* (CUP, Cambridge, 2019), at 147–173.

[18] Notably, the 1999 Protocol on Water and Health to the 1992 UNECE Convention on the Protection of Transboundary Watercourses and International Lakes (17 March 1992, entered into force 6 October 1996) 1936 UNTS 269. See further, A. Tanzi, 'Reducing the Gap between International Water Law and Human Rights Law: The UNECE Protocol on Water and Health', (2010) 12 *International Community Law Review* 267; O. McIntyre, 'The UNECE Water Convention and the Human Right to Access to Water: The Protocol on Water and Health', in A. Tanzi *et al* (eds.), *The UNECE Convention on the Protection and Use of Transboundary Watercourses and International Lakes: Its Contribution to International Water Cooperation* (Brill Nijhoff, Leiden, 2015), at 345–366.

Target 6.1 is also supported by the core values inherent to international water resources law, as largely codified in, and exemplified by, the 1997 UN Watercourses Convention, under which the cardinal rule of equitable and reasonable utilisation dictates that an equitable inter-State allocation of uses or benefits of an international watercourse, or of quantum share of water therefrom, is to be determined having regard to a range of relevant factors which help to characterise each State's economic, social, and environmental dependence upon the shared waters in question. While the UN Watercourses Convention makes it quite clear that, unless watercourse States have agreed otherwise, 'no use of an international watercourse enjoys inherent priority over other uses',[19] it also provides that 'in the event of a conflict between uses of an international watercourse, it shall be resolved ... with special regard being given to the requirements of *vital human needs*'.[20] This long-standing elevation in general international water law of the use of water for meeting vital human needs appears to pre-empt the current discourse on the human right to water and merely reflects long established State and judicial practice confirming that certain basic needs of the population dependent upon shared waters, in particular the use of water for drinking and other domestic purposes, will be accorded priority.[21] The scope of this prioritised use of shared water resources would appear to correspond closely with target 6.1 and the related human right to water. A Statement of Understanding on Article 10(2) was included in the Report of the UN General Assembly Working Group which finalised the drafting of the Convention, providing that, 'in determining "vital human needs", special attention is to be paid to providing sufficient water to sustain human life, including both drinking water and water required for production of food in order to prevent starvation'.[22] Leading commentators explain the legal significance and implications of Article 10(2) in the following terms:

> The protection of vital human needs entails a 'presumptive' priority over all the other factors listed in Article 6, such presumption being rebuttable only on the basis of the specific circumstances of the individual case. That is to say

---

[19] UN Watercourses Convention, art 10(1).
[20] *Ibid.*, art 10(2), (emphasis added).
[21] See, for example, J. Lipper, 'Equitable Utilization', in A.H. Garretson *et al* (eds.), *The Law of International Drainage Basins* (Oceana, New York, 1967), at 60–62.
[22] UNGA, oral report of the coordinator of the informal consultations (Summary Record of the Fifty-Seventh Meeting) on the Convention on the

that watercourse States, in discussing the equitable allocation of shared watercourses, cannot avoid starting negotiations, taking the water supplies needed to support vital human needs as a fixed parameter.[23]

Thus, though Article 6(1) of the UN Watercourses Convention also includes among the 'factors relevant to equitable and reasonable utilization ... the population dependent on the watercourse in each watercourse State', such dependence, at least to the extent that it relates to vital human needs, appears less a factor to be considered and weighed and more a non-negotiable responsibility to be respected and fulfilled by all watercourse States. Indirectly emphasising the priority accorded to such vital human needs by Article 10(2), the International Law Commission (ILC) in its commentary to Article 7 (on the duty to prevent significant transboundary harm) makes it quite clear that 'a use that causes significant harm to human health and safety is understood to be inherently inequitable and unreasonable'.[24] Clearly, Article 10(2) of the Convention is intended to be consistent with the aim of ensuring to all peoples access to safe and sufficient water supplies, as originally set out in Chapter 18 of Agenda 21,[25] further elaborated in General Comment No 15, and now enshrined in SDG target 6.1. Equally, the universal adoption of SDG 6 might do much to make this presumption of priority accorded to the requirements of vital human needs irrebuttable in any circumstances and their safeguarding even more of a *sine qua non* in international water resources law, quite apart from any corresponding obligation arising under international human rights law.

The pre-existing commitment in international water law to the aims of target 6.1 would only appear to have strengthened in recent years, notably with the adoption of such instruments as the 1999 Protocol on Water and Health to the 1992 UNECE Water Convention. This is reflected by

---

Law of the Non-Navigational Uses of International Watercourses (1997) UN Doc A/C.6/51/SR.57, art 10(2).

[23] A. Tanzi and M. Arcari, *The United Nations Convention on the Law of International Watercourses* (Kluwer Law International, Dordrecht, 2001), at 141.

[24] International Law Commission, *Report of the International Law Commission on the Work of its Forty-Sixth Session*,(1994) II *Yearbook of the International Law Commission* A/CN.4/SER.A/1994/Add.1 104, at para 14.

[25] UN Conference on Environment and Development (UNCED), Rio Declaration on Environment and Development, (1992) 31 *International Legal Materials* 874, UN Doc A/CONF.151/26 (vol II).

the 2004 International Law Association (ILA) Berlin Rules on Water Resources,[26] intended to update and replace the seminally important 1966 ILA Helsinki Rules,[27] which appear to afford clear and formal priority to vital human needs, defined under that instrument to mean 'waters used for immediate human survival, including drinking, cooking and sanitary needs, as well as water needed for the immediate sustenance of a household'.[28] Article 14 of the Berlin Rules builds upon Article 10(2) of the UN Watercourses Convention to accord even greater emphasis and priority to the provision of adequate basic water for all by providing that:

1. In determining an equitable and reasonable use, States first allocate waters to satisfy vital human needs.
2. No other use or category of uses shall have an inherent preference over any other use or category of uses.

The Berlin Rules further include a distinct Chapter IV on the 'Rights of Persons', including a dedicated Article 17 on the 'The Right of Access to Water', which provides, inter alia, that 'every individual has a right of access to sufficient, safe, acceptable, physically accessible, and affordable water to meet that individual's vital human needs'.[29] Such provisions clearly reflect a growing recognition in general international water law of a human right of access to water, and of the closely related aspirations set out under SDG 6, which are to be pursued in the case of transboundary water resources by prioritisation of the need to satisfy vital human needs. Thus, while the universal adoption of SDG 6 serves to enhance the legitimacy and significance of the increasingly firmly established human right to water paradigm in international water law, the primacy long afforded to vital human needs in international water law lends support in State practice to human rights-based approaches and entitlements and, thereby, should assist realisation of SDG 6.

---

[26] International Law Association, *Report of the Seventy-First Conference of the International Law Association* (2004) ('Berlin Rules') 337.
[27] International Law Association, Helsinki Rules on the Uses of the Waters of International Rivers, ILA, *Report of the Fifty-Second Conference* (Helsinki 1966) 484.
[28] *Ibid.*, art 3(20).
[29] Berlin Rules on Water Resources, art 17(1).

9.3 SDG target 6.3: 'by 2030, improve water quality by reducing pollution'

The pollution-related objectives included under target 6.3 simply reiterate and strengthen the central importance long accorded to environmental values in modern international water law, as reflected by the inclusion of the detailed and imperative pollution provisions contained in Part IV of the UN Watercourses Convention and the even more exacting provisions of the 1992 UNECE Water Convention, which focus on the prevention, control and reduction of 'transboundary impact'.[30] These instruments reflect long-standing concerns among watercourse States regarding the need to protect watercourses as fragile natural resources providing a range of indispensable benefits, including the provision of drinking water and the removal of wastes, which need to be carefully managed in order to achieve balance and sustainability. Pollution control has long been a central concern of international water law, and in an attempt to codify pollution-related measures, Article 21(2) of the UN Watercourses Convention provides that:

> Watercourse States shall, individually and, where appropriate, jointly, prevent, reduce and control the pollution of an international watercourse that may cause significant harm to other watercourse States or their environment, including harm to human health or safety, to the use of the waters for any beneficial purpose or to the living resources of the watercourse. Watercourse States shall take steps to harmonize their policies in this connection.

Article 21(3) in turn makes provision for the holding of consultations amongst States sharing an international watercourse with a view to arriving at mutually agreeable measures and methods to prevent, reduce and control pollution of an international watercourse, such as:

(a) Setting joint water quality objectives and criteria;
(b) Establishing techniques and practices to address pollution from point and non-point sources;
(c) Establishing lists of substances the introduction of which into the waters of an international watercourse is to be prohibited, limited, investigated, or monitored.

---

[30] UN Watercourses Convention, arts 20–23 and UNECE Water Convention, art 3.

Clearly, this Article, which codifies related practice and typifies the pollution management provisions contained in almost all international watercourse and water resources agreements, corresponds very closely with the aims set out in target 6.3 and the indicators developed thereunder to increase treatment of household wastewater[31] and industrial wastewater[32] and to improve ambient water quality.[33]

In elaborating upon the significance of Article 21, the ILC, in its detailed commentary to the 1994 Draft Articles, appears to regard the due diligence obligation set out under Article 21(2) to prevent, reduce and control pollution as 'applying the general obligation of Article 7 [on prevention of significant transboundary harm] to the case of pollution'. Therefore, in order to come within the scope of Article 21, the pollution in question must be of a type that 'may cause significant harm to other watercourse States or to their environment', of a type corresponding with the illustrative examples listed under Article 21(2), including 'harm to human health or safety, to the use of the waters for any beneficial purpose or to the living resources of the watercourse'. International water law's focus upon pollution likely to give rise to such harm further illustrates its direct coherence with the broad aims of target 6.3 and, more indirectly, with those of targets 6.1 and 6.2. The ILC commentary also confirms that 'the principle of precautionary action is applicable, especially in respect of dangerous substances such as those that are toxic, persistent or bio-accumulative', and recognises that the pollution control framework provided by Article 21 is supported by a wealth of 'representative illustrations of international agreements, the work of international organizations, decisions of international courts and tribunals, and other instances of State practice'. Thus the ILC appears expressly to recognise the critical role of the extensive body of developed practice existing in the field of international environmental law in supporting the effective implementation and application of the key pollution control obligation set out in Article 21 of the UN Watercourses Convention and, by extension, the aims of target 6.3. Of course, State practice developing in the light of the emerging human right to water,

---

[31] Indicator 6.3.1 measures the 'Proportion of household wastewater flow safely treated'.
[32] Indicator 6.3.2 measures the 'Proportion of industrial wastewater flow safely treated'.
[33] Indicator 6.3.3 measures the 'Proportion of bodies of water with good ambient water quality'.

consistent with the stipulations of General Comment No 15, as well as practice now developing in the light of the efforts of States to implement SDG 6, might also be expected to inform aspects of the due diligence obligation to prevent or reduce environmental pollution arising under international water law.

It is only logical that any water-related goal should encompass entitlements and obligations relating to protection of the natural environment, as this will play a pivotal role in the provision of water-related services. General Comment No 15 similarly declares that the right to water involves both the right of the individual to be free 'from unsafe and toxic water conditions' and from 'contamination of water supplies', and the obligation of States to refrain from 'unlawfully diminishing or polluting water' and to adopt legislative and other measures to restrain third parties from polluting water sources. As regards the environmental standards to be applied pursuant to the right to water, General Comment No 15 expressly lists the United Nations Environment Programme (UNEP) among the organisations from whom 'States parties may obtain guidance on appropriate indicators', thereby stressing the direct relevance of UNEP's work on water and freshwater ecosystems. This connection once again highlights the strong interlinkages between the right to water and SDG 6, just as UNEP's Freshwater Strategy 2017–21 focuses on those SDG targets that relate to water quality and pollution, freshwater ecosystems, integrated water resources management (IWRM), and water-related conflict and disasters.[34]

Official guidance on implementation of SDG 6 emphasises the use of existing environmental standards set out under relevant multilateral environmental agreements (MEAs), wherever available. For example, UN-Water guidelines relating to target 6.3 on water quality and wastewater,[35] which aim to protect both ecosystem health and human health by eliminating, minimising, and significantly reducing different streams of pollution into water bodies, advocate doing so in a manner consistent with the Basel Convention on the Control of Transboundary Movements

---

[34] Specifically, SDG targets 6.3, 6.5, 6.6, 11.5 and 16.1. See UN Environment, *Freshwater Strategy 2017–21* (2017), at 6.
[35] UN-Water, *Integrated Monitoring Guide for SDG 6: Targets and Global Indicators* (UN, July 2016), at 8.

of Hazardous Wastes and their Disposal,[36] the Rotterdam Convention on the Prior Informed Consent Procedure for Certain Hazardous Chemicals and Pesticides in International Trade,[37] and the Stockholm Convention on Persistent Organic Pollutants.[38] Regarding standards stipulated thereunder, the latter Convention has established a Persistent Organic Pollutants Review Committee comprising thirty-one experts nominated by the States Parties, which reviews chemicals nominated for listing and control under the Convention having regard to detailed criteria relating to their persistence, bioaccumulation, potential for long-range environmental transport and toxicity (Annex D), to the likelihood of these chemicals leading to significant adverse effects on human health or the environment (Annex E), and to socio-economic considerations associated with possible control measures (Annex F). The same UN-Water guidance also refers to the 1992 UNECE Water Convention and the 1997 UN Watercourses Convention, the former of which is supported by a Secretariat and has been a very important source of technical guidance on all aspects of transboundary water resources management.

Therefore, the wealth of standards and practice developed under the auspices of established MEAs assists in informing the requirements of target 6.3, just as it helps to inform the precise normative implications of the human right to water. The direct and indirect relevance of such a diversity of legal, and even non-legal, environmental standards for understanding the precise requirements of SDG 6 and/or the normative implications of the right to water illustrates the phenomenon of growing normative convergence in international water and environmental law.

---

[36] Basel Convention on the Control of Transboundary Movements of Hazardous Wastes and their Disposal (1989) 1673 UNTS 126.

[37] Rotterdam Convention on the Prior Informed Consent Procedure for Certain Hazardous Chemicals and Pesticides in International Trade (1999) 2244 UNTS 337.

[38] Stockholm Convention on Persistent Organic Pollutants (2001) 2256 UNTS 119; 40.

9.4 SDG target 6.5: 'implement integrated water resources management at all levels, including through transboundary cooperation'

While 'integrated water resources management', as an overarching governance approach for the sustainable management of water and related resources at the basin-level, or as a technical best practice standard in water resources management, enjoys some support in international declaratory practice,[39] its legal status remains unclear and its legal relevance questionable. However, the commitment to transboundary cooperation set out in target 6.5 resonates very clearly with international water law and the subsequent adoption of SDG indicator 6.5.2, which tracks the percentage of the total transboundary basin area within a country that has an operational arrangement for water cooperation, and on which States must report regularly, is potentially game-changing. For the purposes of indicator 6.5.2, an arrangement for water cooperation is a bilateral or multilateral treaty, convention, agreement or other formal arrangement between riparian States that provides a framework for cooperation and any such arrangement will only qualify if it meets each of the following four criteria:

- Existence of a joint body;
- Regular, formal communication between riparian countries (at least once a year);
- Joint or coordinated management plans or objectives; and
- Regular exchange of data and information (at least once a year).

Thus, target 6.5 now provides co-basin States with an unprecedented incentive to establish and participate actively in common management institutions for transboundary basins, along with an outline of the minimum requirements for such cooperative institutions.

One of the ways in which the values set out in the SDGs cohere most markedly with those central to international water law is through the for-

---

[39] See, for example, UNCED, Rio Declaration on Environment and Development, Agenda 21: A Programme for Action for Sustainable Development, *Report of the United National Conference on Environment and Development*, (1992) 31 *International Legal Materials* 874 UN Doc A/CONF.151/26 (vol II), Annex II, at paras 18.5 and 18.6.

mer's explicit advocacy of transboundary cooperation in the governance of water resources, arguably the ultimate, defining aim of the latter body of rules. A general duty upon States to cooperate over transboundary waters is included in almost all modern water resources conventions and declaratory instruments.[40] In addition to requiring watercourse States to cooperate in good faith 'in order to attain optimal utilization and adequate protection of an international watercourse', Article 8 of the UN Watercourses Convention encourages them to 'consider the establishment of joint mechanisms or commissions ... to facilitate cooperation on relevant measures and procedures'. The 1994 ILC commentary on Draft Article 8 leaves little room for doubt as to the legal status or significance of the duty to cooperate. Concluding that transboundary water cooperation provides an important basis 'in order to fulfil the obligations and attain the objectives set forth in the draft articles', the ILC points out that Draft Article 8 'refers to the most fundamental principles upon which cooperation between watercourse States is founded', namely sovereign equality, territorial integrity, mutual benefit, and good faith, before summarising the very wide range of international conventional instruments, declarations, and resolutions calling for cooperation over the utilisation of international watercourses. The duty to cooperate can largely be understood as an 'umbrella' or composite obligation, largely consisting of a range of procedural requirements, any one or more of which might be applicable in a given situation. These notably include the duty to exchange information relevant to use of the watercourse, the duty to notify co-riparian States of planned projects potentially impacting a shared watercourse and, where necessary, duties to consult and negotiate with such States in a good faith effort to address their concerns.[41]

Beyond the general duty to cooperate over the utilisation and protection of shared water resources owed to co-riparian States, General Comment No. 15 clearly identifies a duty on States to cooperate in furthering realisation of the human right to water. Though it notes that the UN Watercourses Convention

> requires that social and human needs be taken into account in determining the equitable utilization of watercourses, that State parties take measures to

---

[40] See generally, C. Leb, *Cooperation in the Law of Transboundary Water Resources* (CUP, Cambridge, 2013).
[41] As exemplified in arts 9 and 11–19 of the UN Watercourses Convention.

prevent significant harm being caused, and [that] special regard must be given to the requirements of vital human needs,

General Comment No 15 further notes, in the context of the international obligations imposed upon States by virtue of Articles 11 and 12 of the 1966 International Covenant on Economic, Social and Cultural Rights, that

> To comply with their international obligations in relation to the right to water, States parties have to respect the enjoyment of the right in other countries. *International cooperation requires* States parties to refrain from actions that interfere, directly or indirectly, with the enjoyment of the right to water in other countries. Any activities undertaken within the State party's jurisdiction should not deprive another country of the ability to realize the right to water for persons in its jurisdiction (emphasis added).

In fact, the CESCR goes so far as to include the duty to cooperate in realising the right to water as one of the 'core obligations' of States Parties to the 1966 Covenant. Thus, the international discourse on the human right to water had already begun to alter the focus of the fundamental duty to cooperate under international water law towards pursuit of the availability and sustainable management of water and sanitation for all, an evolutionary transformation supported in no small measure by the successive universal adoption by States of MDG 7 and SDG 6.

It should be noted that General Comment No 15 is emphatic in linking good faith transboundary water cooperation, as required under international water law, to the realisation of the human right to water in a number of ways. For example, in listing indicative illustrations of violations of the right to water, it includes the 'failure of a State to take into account its international legal obligations regarding the right to water when entering into agreements with other States or with international organizations'. General Comment No 15 further provides that 'depending on the availability of resources, States should facilitate realization of the right to water in other countries, for example through provision of water resources'. Though this provision is couched in soft terms, it could clearly be understood to recognise a requirement that, in utilising shared water resources, States must ensure that adequate water is available for the realisation of the human right to water in co-basin States, once again linking human rights requirements to consideration of vital human needs in the practice of inter-State water resources allocation. Likewise, General Comment No 15 generally suggests that

> States parties should ensure that the right to water is given due attention in international agreements and, to that end, should consider the development of further legal instruments. With regard to the conclusion and implementation of other international and regional agreements, States parties should take steps to ensure that these instruments do not adversely impact upon the right to water.

Clearly, this statement suggests that States should only conclude global or regional water resources conventions and river basin treaties that are compatible with full realisation of the human right to water in all co-basin States and should act to further develop and adapt existing treaty regimes to ensure such compatibility. All this suggests that the CESCR envisages a particularly important role for international water law in the realisation of the human right to water and, consequently, for effective realisation of SDG 6.

Demonstrating the integrated and indivisible nature of the 2030 Agenda, it is possible to link target 6.5 to SDG 16, suggesting that the former should be analysed in conjunction with the latter, which is concerned with strengthening international cooperation and transboundary good governance. Clearly this aim resonates with the general objectives of the UN Watercourses Convention and of General Comment No 15. For example, target 16.3 seeks to 'promote the rule of law at the national and international levels, and ensure equal access to justice for all' which, in addition to promoting adherence to established rules of international water law, would enable individuals to react to breaches of both substantive and procedural obligations of international water law. Though international water law has not traditionally concerned itself with the rights of individuals, Article 32 of the UN Watercourses Convention, on 'non-discrimination' regarding access to legal redress 'for the protection of the interests of persons, natural or juridical, who have suffered or are under a serious threat of suffering significant transboundary harm as a result of activities related to an international watercourse', appears to anticipate a role for private recourse by adversely affected persons to domestic courts and remedies as a means for establishing effective responsibility for unlawful activities related to international watercourses.[42]

Generally, international water law appears to be developing in a manner consistent with the 'transboundary good governance' values embraced by SDG 16, such as target 16.6, which calls upon States to 'develop effective, accountable and transparent institutions at all levels', target 16.7, which seeks to 'ensure responsive, inclusive, participatory and representative decision-making at all levels', and target 16.10, which exhorts States to 'ensure public access to information and protect fundamental freedoms in accordance with national legislation and international agreements'. Regarding institutions, for example, transboundary water cooperation very commonly involves the establishment of inter-State institutional machinery to formulate and implement common policies for the management and development of the basin,[43] an approach to managing shared water resources widely endorsed by the international community, and by international bodies concerned with codification of this field of international law.

While the UNECE Water Convention rather unusually imposes a general obligation upon State parties to participate in the establishment of such 'joint bodies',[44] the UN Watercourses Convention merely encourages watercourse States to enter into common management arrangements consistent with the requirements of the principle of equitable and reasonable utilisation, the duty to prevent significant transboundary harm, and the general duty to cooperate.[45] Increasingly, such institutions are subjected to requirements of transparency and accountability and to other rapidly evolving constraints of administrative good governance.

---

[42] Similarly, arts 69–71 of the Berlin Rules also provide for adversely affected legal individuals to seek recourse before a competent judicial or administrative authority in the State where the harm arises.

[43] See generally, S. Schmeier, *Governing International Watercourses: The Contribution of River Basin Organizations to the Effective Governance of Internationally Shared Rivers and Lakes* (Routledge, London, 2013); I. Dombrowsky, *Conflict, Cooperation and Institutions in International Water Management: An Economic Analysis* (Edward Elgar, Cheltenham, 2007); O. McIntyre, 'The Legal Role and Context of River Basin Organisations', in A. Kittikhoun and S. Schmeier (eds.), *Water Diplomacy and Conflict Management* (Routledge, London, 2021).

[44] UNECE Water Convention, art 9(2).

[45] UN Watercourses Convention, arts 5(2), 8(2), 9, 21, 24, and 33(2).

Regarding responsive decision-making and public access to information, the practice of modern international water law has long revolved around processes of prior inter-State notification of planned measures and routine information exchange on the state of shared water resources. However, in recent years such processes have begun to open up in order to facilitate or require broader engagement with the public or, at least with key stakeholders.[46] This can be seen in the 2010 International Court of Justice (ICJ) judgment in the *Pulp Mills* case, where the Court placed great emphasis on the significance of procedural requirements, and in particular the duty to conduct environmental impact assessment (EIA), in discharging the duty to cooperate over planned projects, while also stressing the role of formal institutional machinery established to facilitate meaningful cooperative engagement.[47] Of course, any credible EIA process will involve significant public and stakeholder disclosure, consultation and engagement. It is clear, therefore, that international water law can and must adapt in order to contribute to realisation of the 'transboundary good governance elements' of the SDGs, but equally that the SDGs have a key role in driving further progressive development of this body of law.

9.5 SDG target 6.6: 'protect and restore water-related ecosystems'

The inclusion of a dedicated global target and indicator[48] on the protection and restoration of water-related ecosystems reflects the consensus evident in modern international water law regarding the critical importance of maintaining such ecosystems for water resources management, as well as the central role of water in maintaining a range of vital ecosystem services. This consensus is embodied in the ecosystem protection obligation set out in Article 20 of the UN Watercourses Convention. The 'ecosystem approach' is understood as a management approach which can inform the

---

[46] For example, Agreement on the Establishment of the Zambezi Water Commission (Kasane, 13 July 2004), art 16(8).

[47] *Pulp Mills on the River Uruguay (Argentina v Uruguay)* (Judgment) [2010] *ICJ Reports* 14, paras 77 and 87–91.

[48] Indicator 6.6.1 measures the 'proportion of river basins showing high surface water extent changes'.

application of legal frameworks for water resources utilisation, land-use planning, and development control at both the national and transboundary levels. In the specific context of international water law, the concept has been linked to the obligation to protect international watercourses ancillary to the principle of equitable and reasonable utilisation, and accordingly to the objective of sustainable development. The evolution in scientific understanding which has given rise to the ecosystem approach continues and is increasingly reflected in the emergence of sophisticated methodologies which function to inform the normative implications of State obligations to protect and sustainably manage watercourse ecosystems, including the maintenance of environmental flow regimes and safeguarding of ecosystem services. Of course, lessons learned in the implementation of legal requirements arising under other environmental conventions to protect and preserve associated water-related ecosystems may also inform the environmental and ecosystem obligations set out under Part IV of the UN Watercourses Convention.

Whereas formal recognition in the UN Watercourses Convention of the normative character and significance of ecosystem protection requirements merely reflects the continuing evolution of international environmental law more generally, notably under the Convention on Biological Diversity (CBD),[49] the universal adoption of target 6.6 amounts to a very significant endorsement to what is rapidly emerging as a key value in transboundary water resources management. On the other hand, the increasingly sophisticated legal and technical requirements for ecosystems protection emerging in international water law provide a vital tool for realising target 6.6 in the important context of international basins. It is increasingly clear that continuing elaboration and development of the ecosystem approach to the management of water resources, and of its constitutive normative elements, will be critically important for effective realisation of the new overarching imperative of international water law, that is, the optimal and ecologically sustainable use of shared water resources in an era of looming freshwater scarcity. Quite apart from inexorably rising demand for water, food, and energy, and associated large-scale water resources utilisation, international water law will have to contend with the very significant ecological challenges posed by climate change and the adaptation measures required to address it.

---

[49] In addition, CBD art 8(f) obliges state parties to 'rehabilitate and restore degraded ecosystems'.

Thus, the broad social and environmental objectives of SDG 6, and the practice of States that emerges thereunder, may now inform every aspect of international environmental and natural resources law, which can function to assist attainment of these same objectives. In articulating the cardinal obligation of watercourse states to 'utilize an international watercourse in an equitable and reasonable manner', the UN Watercourses Convention stresses that they shall do so 'with a view to attaining optimal and sustainable utilization thereof and benefits therefrom ... consistent with adequate protection of the watercourse'.[50] The factors identified as relevant to this principle expressly include 'ecological and other factors of a natural character'[51] and '[c]onservation, protection, development and economy of use of the water resources of the watercourse'.[52] The duty to prevent significant transboundary harm remains central to international water law,[53] but must increasingly be understood to include novel forms of ecological disturbance in the light of heightened environmental sensibilities and the wealth of technical guidance on ecological standards and ecosystems management produced under multilateral environmental agreements. The 1994 ILC commentary hints at the reasoning behind the priority afforded to ecosystems protection, explaining that 'protection and preservation of aquatic ecosystems help to ensure their continued viability as life support systems, thus providing an essential basis for sustainable development'.

9.6 Conclusion

The global initiative represented by the adoption and progressive implementation of SDG 6 represents an innovative and potentially game-changing development in terms of the equitable and sustainable management of water resources and related ecosystems and services, particularly in the case of watercourses shared between two or more States. This is not before time, having regard to the increasingly urgent nature of the problems facing transboundary water management and the often, poor results achieved by reliance solely upon traditional

---

[50] UN Watercourses Convention, art 5(1).
[51] *Ibid.*, art 6(1)(a).
[52] *Ibid.*, art 6(1)(f).
[53] *Ibid.*, art 7. See also art 6(1)(d).

governance techniques, including the existing formal international legal frameworks. By obtaining the universal voluntary commitment of the international community of States to a set of core water-related values and objectives and presenting these within a structured, coherent, and incremental governance framework that includes periodic review, annual reporting, and centralised monitoring, SDG 6 provides the shared vision to guide global water management and governance for the foreseeable future. From this perspective, SDG 6 represents, by any measure, a truly extraordinary achievement on the part of the institutional machinery of global cooperation.

However, this success is due in large measure to the fact that those involved in negotiating and framing SDG 6 did not have to commence the task *de novo*. In addition to the valuable experience gained in implementing the preceding water-related objectives under target C of MDG 7, they could draw upon the wealth of ongoing progressive developments in the fields of international water, human rights and environmental law, all of which can be associated with the broad objective of sustainable development. The carefully formulated substantive and procedural entitlements included within the rubric of the human right of access to water, the established legal frameworks for balancing the social, economic and developmental water needs of competing co-riparian States, the emerging ecosystem-based approach to the management of international water resources and the ecosystem services provided thereby, and the increasingly ubiquitous participation rights granted to water users, other stakeholders, and the general public, will continue to inform implementation of every aspect of SDG 6. In many respects this global goal for water consolidates existing trends in relevant international legal frameworks. Such influence runs in both directions, however, and, as the overarching global governance framework for water, the values and methodologies set out under SDG 6 will exert ever-increasing influence over the continuing development of international water law frameworks and their application. Inter-State engagement regarding shared water resources already has careful regard to the requirements of SDG indicator 6.5.2.[54] The influence of SDG 6 as a key source of normative guidance as regards transboundary water governance can only continue to grow. Thus SDG 6 and international water-related legal frameworks enjoy a highly synergis-

---

[54] See UN-Water, *Progress on Transboundary Water Cooperation: Global Status of SDG Indicator 6.5.2 and Acceleration Needs* (UN, Geneva, 2021).

tic and complementary relationship. SDG 6 provides a formally adopted, yet voluntary global consensus on a comprehensive range of water-related values, along with elaborate compliance monitoring processes, while traditional international water law provides a formal framework of notionally binding rules which are often opaque as regard their precise requirements and/or lack practically effective enforcement mechanisms. Each framework can only function to assist realisation of the other.

Index

1648 Peace of Westphalia 14, 15
1792 French Decree 15, 137
1815 Vienna Congress 137, 142
1919 Treaty of Versailles 15
1966 International Covenant on
    Economic, Social and Cultural
    Rights 179
    Articles 11 and 12 190
1992 Agreement between Namibia
    and South Africa on the
    Establishment of a Permanent
    Water Commission
    Article 1(2) 138
1999 Protocol on Water and Health
    182
2000 Revised SADC Protocol on
    Shared Watercourses 138
2030 Agenda 176–9, 191

absolute territorial integrity 5, 31–2
absolute territorial sovereignty 4, 31
abuse of rights doctrine 71
*Action Plan for the Human
    Environment* 33
    Recommendation 51 49, 135–6
adaptive management 26–7
    conventional frameworks and 148
    described 147–8
    EIA 148
    legal frameworks
        precautionary principle 123
        traditional 122–3
        for transboundary
            cooperation 123–4
    precautionary principle 149
    resilience of ecosystem 122

transboundary basins 148
Agenda 21: Chapter 18 182
Agreement between Namibia
    and South Africa on the
    Establishment of a Permanent
    Water Commission (1992)
    see 1992 Agreement between
    Namibia and South Africa
    on the Establishment
    of a Permanent Water
    Commission
Alabama Claims Arbitration (1872) 74
Anzilotti (Judge) 44

Babylonian Code of Hammurabi 14
Basel Convention on the Control of
    Transboundary Movements of
    Hazardous Wastes and their
    Disposal 187
benefit-sharing
    arrangements 29–30, 151–2
    described 55–6
    legal basis 58–62
    practice to international water
        resources 56–8
Berlin Rules on Water Resources
    (2004) 19, 23, 28, 86
    Article 14 183
    Article 64 139
    Chapter IV on the 'Rights of
        Persons' 183
    Chapter VII on Extreme
        Situations 108
    Chapter XI 96
Boundary Waters Treaty (1909) 57,
    143

broadly conceived equity approach 52
Brownlie, I. 47

Canelas de Castro, P. 170
CBD Conference of the Parties (COP) 23
CBD COP 5 125
CBD-ification 156
CBD Strategic Plan for Biodiversity (2011–20) 121
CESCR *see* UN Committee on Economic Social and Cultural Rights (CESCR)
Chapter 18 of Agenda 21 136
cognitive learning 156
Columbia River Treaty (1961) 56–7
common but differentiated responsibilities (CBDR) principle 65–8, 87
common heritage equity 52–3
common management approach 7–8, 50, 135–6
common management institutions
  common management arrangements 139–41
  described 139
  effectiveness of machinery 142
  equitable participation principle 140
  equitable and reasonable utilisation 141–2
  prevention, reduction and control of pollution 141
community of interest 7–8, 67, 136–8
concept of harm
  ecosystem approach 73–4
  Helsinki Rules: Article X 72–3
conceptual evolution of international water law
  absolute territorial integrity 5
  absolute territorial sovereignty 4
  common management approach 7–8
  community of interest 7–8
  limited territorial sovereignty 5–6
  SDG 6 10–11
consultation and negotiation 104–7
contemporaneity principle 168

Convention on Biological Diversity (CBD) 23, 97–8, 121, 146, 156, 194
  Articles 1 and 15(7) 39
Convention on Environmental Impact Assessment in a Transboundary Context
  Article 2(1) 82
Convention on the Law of the Non-navigational Uses of International Watercourses
  Article 6 78
convergence 8–9, 170–71
  drivers of 160
    ICJ *see* International Court of Justice (ICJ)
    ILC *see* International Law Commission (ILC)
    International Human Rights Law 163
  forms of 159
  in international law
    fragmentation and 156–7
    ILC 157–8
    systemic integration principle 158
  and international water law 163–4
    basis for responsibility 169
    divergence and fragmentation 165–6
    dynamism of modern international water law 164–5
    ecological values 167
    ecosystems obligations 170
    EIA 169–70
    foundational cases 168
    human rights values 167
    rules of hierarchy 166–7
    SDG 6 167
    substantive fragmentation 169–70
    UNECE Water Convention 168–9
cooperation and procedural rules 91–4, 109–10
  EIA 93–4

procedural due diligence
 obligation 91–2
 substantive due diligence 91–2
corrective equity 52
Craven, M. 166
Crawford, J. 162, 163
customary international law 19

Danube Commission 142, 143
Danube Convention (1994) 142
de minimis rule 73
distributive equity 42, 65–6
doctrine of 'prior appropriation' 46
Draft Articles on the Law of the
 Non-Navigational Uses of
 International Watercourses 18
Draft Articles on Transboundary
 Aquifers 2
due diligence 130–31
 context-specific standards
  MEA 82
  no-harm rule 81
  primary rules to shared water
   resources 82–3
 described 74
 ecological 84–6
 procedural 84
 standard of care 75
 standards, generally applicable
  2–3
 Draft Article 3 on Prevention
  80
 Draft Article 10 on
  Prevention 78–9
 duty to prevent 76, 79
 environmental context 80
 good government 77
 precautionary principle 79
 Principle 3 of Rio
  Declaration 81
 reasonableness 77
 risk of significant
  transboundary harm
  78
 State of origin 77–8
 substantive 74, 84
duty to consult/negotiate in good faith
 104–7

duty to notify
 failure to respond to notification
  99
 inter alia 97, 99
 issues 98
 obligation to refrain 100–101
 Watercourses Convention details
  in relation to 101–2
duty to warn 107–9

ecological due diligence 84–6
ecological sustainability 19–23
 ecosystems
  approach 20–21
  defined 21
  obligation to conserve 23
  protection 21–2
 pollution control 19
 UNECE Water Convention 22–3
ecosystems 21–2
 approach *see* ecosystems
  approach
 defined 21
 obligation to conserve 23
 protection 21–2
 services *see* ecosystem services
ecosystems approach 11, 20–21, 73–4,
 85, 113, 117–18, 126–7, 140,
 143–4
 adaptive management *see*
  adaptive management
 benefit-sharing arrangements
  151–2
 ecosystem services *see* ecosystem
  services
 environmental flows *see*
  environmental flows
 stakeholder and public
  participation 149–51
ecosystem services 25–6, 29, 133
 categories 120–21
 concept described 145–6
 ecosystem components and 121
 legal framework 121–2
 PES 147, 152
 in transboundary water
  cooperation 146–7
ecosystems protection 115–17

EIA *see* environmental impact assessment (EIA)
emergency, defined 109
environmental flows 24–5, 144–5
　defined 119, 144
　described 118–19
　guiding elements 120
　legal obligation 145
　legal significance of maintaining flow 119–20
　minimum 144–5
environmental harm 88–9
environmental impact assessment (EIA) 27, 125, 169–70
　adaptive management 148
　convergence 169–70
　cooperation and procedural rules 93–4
　stakeholder and public participation 125, 150–51
environmental protection 111–13, 133
　adaptive management 122–4
　ecosystem approach *see* ecosystems approach
　ecosystem services *see* ecosystem services
　ecosystems protection 115–17
　environmental flows *see* environmental flows
　international river commissions and 142–52
　pollution control 113–15
　stakeholder and public participation 124–6
equitable and reasonable utilisation 5–6, 31–4, 69
　benefit-sharing *see* benefit-sharing
　equity *see* equity
　legal status 34–5
　normative content
　　transboundary water resources 38
　　UN Watercourses Convention 35–8
　relationship with 86–8
equitable participation principle 140
equity
　broadly conceived 52
　common heritage 52–3
　corrective 52
　as general principle of law
　　River Meuse case 44–6
　　sustainable development 47–8
　international environmental law 39–41
　international human rights law 41–2
　international water resources law 42–55
　procedural 48–50
　and proportionality 128–9
　as proportionality 53–5
　solidarity and 63–8
　as substantive rule of apportionment 50–53
　　broadly conceived equity 52
　　common heritage equity 52–3
　　corrective equity 52
　　Libya-Malta case 51
　　Tunisia-Libya Continental Shelf case 51–2
equity ex aequo et bono 43
erga omnes doctrine 161
exchange of information 102–4

Farraka Treaty (1996) 123
fragmentation 8–9
Franck, T. 35, 52, 142
French Decree (1792) 15, 137
freshwater crisis 1

Gabcíkovo-Nagymaros case 34, 95–6
general duty of cooperation 94–5
　duty of States to cooperate 95–6
　general obligation to cooperate 96
　material harm 97
good faith
　duty to consult/negotiate in 104–7
good government 77
good neighbourliness doctrine 71

Harmon 31, 32
Harmon Doctrine 4, 5, 31
Helsinki Rules (1966) 2, 3, 8, 17, 18, 37, 51, 59, 71, 111, 160, 183
    Article X 72–3
Higgins, R. 51
historical development
    age-old concern 14–15
    navigation to utilisation, from
        codification of international rules 16–17
        customary international law 19
        economic uses of international watercourses 17–18
        industrialisation of Europe and North America 16
        non-navigational uses of international watercourses 18
Hudson (Judge) 44–6
hydro-solidarity 64–5

ICJ *see* International Court of Justice (ICJ)
ICPDR *see* International Commission for the Protection of the Danube River (ICPDR)
IJC *see* US-Canada International Joint Commission (IJC)
ILA *see* International Law Association (ILA)
ILC *see* International Law Commission (ILC)
Incomati-Maputo Agreement (2002) 123
Indus Waters Treaty (1960)
    Article 6 102
influence of international environmental law
    due diligence 130–31
    equity and proportionality 128–9
    invasive species 131–2
    marine environment 132–3
    MEAs 126–8
    precaution 129–30
institutional cooperation 7, 153

see also common management institutions
Integrated Water Resources Management (IWRM) Plan 166, 186
inter alia 97, 99, 141
International Boundary and Water Commission 142
International Columbia River Engineering Board 57
International Commission for the Protection of the Danube River (ICPDR) 166
international community 162
International Court of Justice (ICJ) 154, 160–62
International Covenant on Economic, Social and Cultural Rights (ICESCR) (1966) *see* 1966 International Covenant on Economic, Social and Cultural Rights
international environmental law 155
international human rights law 163
International Joint Commission 57
International Law Association (ILA) 2, 17, 111
    Article V(2) of 35
    Berlin Rules on Water Resources 19
International Law Association Study Group 80
International Law Association Study Group on Due Diligence 77, 84
International Law Commission (ILC) 2, 157–8, 162
    1994 conception of the 'ecological balance' 132
    1994 Draft Articles 71, 113
    2001 Draft Articles on the Prevention of Transboundary Harm from Hazardous Activities 41
    2008 Draft Articles on Transboundary Aquifers 66, 118
    Draft Article 3 on Prevention 77

Draft Articles on Transboundary
        Aquifers 23
International Union for the
        Conservation of Nature
        (IUCN) 24
    environmental flows defined 144
international water institutions 30
invasive species 131–2
IUCN *see* International Union for
        the Conservation of Nature
        (IUCN)

jus cogens doctrine 161

Kishenganga Arbitration 119
Koskenniemi, M. 75

*Lac Lanoux Arbitration (Spain v
        France)* (1957) 5, 91, 95, 104,
        105
law of co-existence 64
law of cooperation 64
limited territorial sovereignty 5–6
Lowe, V. 51

marine environment 132–3
McCaffrey, S. C. 116
MDGs *see* Millennium Development
        Goals (MDGs)
MEAs *see* multilateral environmental
        agreements (MEAs)
Mekong Agreement (1995)
    Article 24(C) 102
Mekong River Commission 121
Meuse and Scheldt Agreements (1994)
        142
Millennium Development Goals
        (MDGs) 172
    SDGs *vs.* 173
Millennium Ecosystem Assessment
        (2005) 120, 146
Ministerial Forum of the Parties
        of the Orange-Senqu River
        Commission (ORASECOM)
        166
Model Provisions on Transboundary
        Groundwaters 2

Multi-Country Cooperation
        Mechanism for the Stampriet
        Transboundary Aquifer System
        (STAS MCCM) 166
multilateral environmental agreements
        (MEAs) 82
    CBD 127
    ecosystem approach 126–7
    wetlands 127

no-harm rule 71–2
    concept of harm
        ecosystem approach 73–4
        Helsinki Rules: Article X
            72–3
    equitable and reasonable
        utilisation, relationship
        with 86–8
Nuclear Tests cases 104

payment for ecosystem services (PES)
        147
PCIJ *see* Permanent Court of
        International Justice (PCIJ)
Peace of Westphalia (1648) 14, 15
Permanent Court of International
        Justice (PCIJ) 2, 3, 7
PES *see* payment for ecosystem
        services (PES)
polluter-pays principle 44
pollution control 113–15
precaution 129–30
precautionary principle 79
    adaptive management 149
Prevention
    Draft Article 3 on 80
    Draft Article 10 on 78–9
prima facie equitable apportionment
        55
procedural due diligence
    described 84
    obligation 91–2
procedural engagement 13, 24, 27, 28,
        33, 96, 110, 126, 139, 142, 151,
        152
procedural rules
    duty to consult/negotiate in good
        faith 104–7

duty to notify 97–102
duty to warn 107–9
and institutional arrangements
  benefit-sharing
    arrangements 29–30
  international water
    institutions 30
  stakeholder and public
    participation 28–9
ongoing exchange of information
  102–4
proportionality and equity 53–5,
  128–9
Protocol on Water and Health (1999)
  182
public participation and stakeholder
  149–51
  basin agreements 124–5
  Convention on Biological
    Biodiversity 125–6
  EIA 125
Pulp Mills case 48, 74

Ramsar Convention 23, 121, 128, 146,
  167
reasonableness 77
Revised 2000 SADC Protocol on
  Shared Watercourses
  Article 3(6) 102
Revised SADC Protocol on Shared
  Watercourses (2000) 138
Rhine Chlorides Arbitration (2004) 67
Rhine Commission 1
Rio Declaration on Environment and
  Development (1992) 65, 155,
  176
  Principle 3 40, 81
  Principle 5 81
  Principle 7 40, 68
  Principle 10 124, 149
  Principle 15 79
  Principle 19 98
river basin organisations (RBOs) 30,
  34, 50, 98, 150, 166
River Meuse case 44–6

Rotterdam Convention on the Prior
  Informed Consent Procedure
  for Certain Hazardous
  Chemicals and Pesticides in
  International Trade 187

Schwebel (Judge) 106
SDG 6 *see* Sustainable Development
  Goal 6 (SDG 6)
SDG 8 68
SDGs *see* Sustainable Development
  Goals (SDGs)
SDG target 6.1: 'by 2030, universal
  access to safe and affordable
  drinking water'
  economic accessibility 180
  General Comment No 15 179–80
  objectives 179
  physical accessibility 180
  UN Watercourses Convention
    181
SDG target 6.3: 'by 2030, improve
  water quality by reducing
  pollution'
  General Comment No 15 186
  UN Watercourses Convention
    Article 21(2) 184, 185
    Article 21(3) 184–5
  UN-Water guidelines 186–7
SDG target 6.5: 'implement integrated
  water resources management
  at all levels, including through
  transboundary cooperation'
  criteria 188
  decision-making and public
    access to information 193
  SDG 16 and 191–2
  UN Watercourses Convention
    Article 8 189
    Article 32 191–2
SDG target 6.6: 'protect and restore
  water-related ecosystems'
  193–5
Seabed Disputes Chamber of the
  International Tribunal of the
  Law of the Sea 81
solidarity 87
  CBDR principle 65–8

community of interest 67–8
    described 63
    equity and 63–8
    hydro-solidarity 64–6
Southern African Development
    Community (SADC) Protocol
    on Shared Watercourse
    Systems (1995)
    Article 1(2) 137
stakeholder and public participation
    28–9, 149–50
    basin agreements 124–5
    Convention on Biological
        Biodiversity 125–6
    EIA 125, 150–51
    RBOs 150
standards of State conduct
    Draft Article 3 on Prevention 80
    Draft Article 10 on Prevention
        78–9
    duty to prevent 76, 79
    environmental context 80
    good government 77
    precautionary principle 79
    Principle 3 of Rio Declaration 81
    reasonableness 77
    risk of significant transboundary
        harm 78
    State of origin 77–8
STAS MCCM *see* Multi-Country
    Cooperation Mechanism for
    the Stampriet Transboundary
    Aquifer System (STAS MCCM)
Stockholm Conference (1972) 49, 136
Stockholm Convention on Persistent
    Organic Pollutants 187
Stockholm Declaration on the Human
    Environment (1972) 70
substantive due diligence 74, 84, 91–2
Sustainable Development Goal 6 (SDG
    6) 10–11, 42, 68, 172–3, 195–7
    2030 Agenda 177–8
    legal frameworks 175–6
    Rio Declaration on Environment
        and Development (1992)
        176–7
    SDG target 6.b 174
    specific targets associated 173–4

target 6.3 *see* SDG target 6.3:
    'by 2030, improve water
    quality by reducing
    pollution'
vital human needs concept 174
water access and management
    176
Sustainable Development Goals
    (SDGs) 10–11, 172–3
    goals and targets 178–9
    MDGs *vs.* 173
    target 6.1 *see* SDG target 6.1: 'by
        2030, universal access
        to safe and affordable
        drinking water'
    target 6.3 *see* SDG target 6.3:
        'by 2030, improve water
        quality by reducing
        pollution'
    target 6.5 *see* SDG target 6.5:
        'implement integrated
        water resources
        management at all
        levels, including
        through transboundary
        cooperation'
    target 6.6 *see* SDG target 6.6:
        'protect and restore
        water-related ecosystems'
systemic integration principle 168

target 6.1 of SDG *see* SDG target 6.1:
    'by 2030, universal access to
    safe and affordable drinking
    water'
target 6.3 of SDG *see* SDG target 6.3:
    'by 2030, improve water quality
    by reducing pollution'
target 6.5 of SDG *see* SDG target 6.5:
    'implement integrated water
    resources management at
    all levels, including through
    transboundary cooperation'
target 16.6 of SDG 192
target 16.7 of SDG 192
target 16.10 of SDG 192
territorial sovereignty 4
Trail Smelter Arbitration 70

INDEX   205

transboundary impact 85
transboundary water cooperation *see* SDG target 6.5: 'implement integrated water resources management at all levels, including through transboundary cooperation'
travaux préparatoires 86
Treaty of Karlstad (1905)
    Article 4 137
Treaty of Versailles (1919) 15, 16

UN Committee on Economic Social and Cultural Rights (CESCR) 179
UN Conference on the Human Environment (UNCHE) (1972) 33
UNECE Water Convention (1992) 2, 3, 7, 70, 99, 182
    Article 2(1) 82
    Article 5(c) 40
    Article 10 105
    Article 13 102–3
    Article 14 108
    ecosystem approach 117–18
UNEP *see* United Nations Environment Programme (UNEP)
UN General Assembly 2
UN General Assembly Resolution (2010) 177
United Nations Conference on the Human Environment (UNCHE) 154
United Nations Economic Commission for Europe (UNECE) Water Convention *see* UNECE Water Convention (1992)
United Nations Environment Programme (UNEP) 39
    Freshwater Strategy 2017–21 186
United Nations Framework Convention on Climate Change (1992)
    Articles 3(1) and 4(2)(a) 39

Universal Declaration of Human Rights 176
UN Special Rapporteur on Human Rights and the Environment 146
UN Watercourses Convention (1997) 2, 3, 7, 18, 21, 28
    Article 4 49
    Article 5 114, 116
    Article 5(1) 55, 58
    Article 5(2) 50
    Article 6 46, 59, 67, 114
    Article 6(1) 35–6, 43, 54, 87, 182
    Article 6(2) 48, 51
    Article 6(3) 36
    Article 7 60, 73, 74, 79, 86, 88
    Article 7(2) 59–60
    Article 8 61–2, 96, 139, 189
    Article 8(2) 62
    Article 9 124, 150
    Article 9(1) 103
    Article 9(3) 103
    Article 10(2) 181–2
    Article 12 99–100, 101
    Article 13 101
    Article 13(b) 49
    Article 14 101
    Article 15 101, 106
    Article 16 101–2
    Article 17 106–7
    Article 18 106, 107
    Article 19 106
    Article 20 128–30, 193–4
    Article 21 83, 129, 185–6
    Article 21(1) 114
    Article 21(2) 114–15, 184
    Article 21(3) 184–5
    Article 22 131–2
    Article 23 132–3
    Articles 11–19 49
    Articles 20–23 73
    General Comment No. 15 189–91
    Part IV 112, 115, 184, 194
US-Canada International Joint Commission (IJC) 143
utilisation
    from navigation to

codification of international
    rules 16–17
customary international
    law 19
economic uses of
    international
    watercourses 17–18
industrialisation of Europe
    and North America 16
non-navigational uses
    of international
    watercourses 18

Vienna Congress (1815) 137, 142
Vienna Convention on the Law of
    Treaties
    Article 33(1)(c) 158

'vital human needs' concept 3, 10, 36,
    67, 167, 174, 181–3

water quality and pollution *see* SDG
    target 6.3: 'by 2030, improve
    water quality by reducing
    pollution'
water-related ecosystems *see* SDG
    target 6.6: 'protect and restore
    water-related ecosystems'
Weeramantry (Judge) 124
Wilde, R. 163
World Trade Organisation (WTO)
    Appellate Body 156
    dispute settlement system 9
WTO *see* World Trade Organisation
    (WTO)

Titles in the **Elgar Advanced Introductions** series include:

International Political Economy
Benjamin J. Cohen

The Austrian School of Economics
Randall G. Holcombe

Cultural Economics
Ruth Towse

Law and Development
Michael J. Trebilcock and Mariana Mota Prado

International Humanitarian Law
Robert Kolb

International Trade Law
Michael J. Trebilcock

Post Keynesian Economics
J.E. King

International Intellectual Property
Susy Frankel and Daniel J. Gervais

Public Management and Administration
Christopher Pollitt

Organised Crime
Leslie Holmes

Nationalism
Liah Greenfeld

Social Policy
Daniel Béland and Rianne Mahon

Globalisation
Jonathan Michie

Entrepreneurial Finance
Hans Landström

International Conflict and Security Law
Nigel D. White

Comparative Constitutional Law
Mark Tushnet

International Human Rights Law
Dinah L. Shelton

Entrepreneurship
Robert D. Hisrich

International Tax Law
Reuven S. Avi-Yonah

Public Policy
B. Guy Peters

The Law of International Organizations
Jan Klabbers

International Environmental Law
Ellen Hey

International Sales Law
Clayton P. Gillette

Corporate Venturing
Robert D. Hisrich

Public Choice
Randall G. Holcombe

Private Law
Jan M. Smits

Consumer Behavior Analysis
Gordon Foxall

Behavioral Economics
John F. Tomer

Cost–Benefit Analysis
Robert J. Brent

Environmental Impact Assessment
Angus Morrison-Saunders

Comparative Constitutional Law,
Second Edition
Mark Tushnet

National Innovation Systems
Cristina Chaminade, Bengt-Åke Lundvall and Shagufta Haneef

Ecological Economics
Matthias Ruth

Private International Law and Procedure
Peter Hay

Freedom of Expression
Mark Tushnet

Law and Globalisation
Jaakko Husa

Regional Innovation Systems
Bjørn T. Asheim, Arne Isaksen and Michaela Trippl

International Political Economy
Second Edition
Benjamin J. Cohen

International Tax Law
Second Edition
Reuven S. Avi-Yonah

Social Innovation
Frank Moulaert and Diana MacCallum

The Creative City
Charles Landry

International Trade Law
Michael J. Trebilcock and Joel Trachtman

European Union Law
Jacques Ziller

Planning Theory
Robert A. Beauregard

Tourism Destination Management
Chris Ryan

International Investment Law
August Reinisch

Sustainable Tourism
David Weaver

Austrian School of Economics
Second Edition
Randall G. Holcombe

U.S. Criminal Procedure
Christopher Slobogin

Platform Economics
Robin Mansell and W. Edward Steinmueller

Public Finance
Vito Tanzi

Feminist Economics
Joyce P. Jacobsen

Human Dignity and Law
James R. May and Erin Daly

Space Law
Frans G. von der Dunk

National Accounting
John M. Hartwick

Legal Research Methods
Ernst Hirsch Ballin

Privacy Law
Megan Richardson

International Human Rights Law
Second Edition
Dinah L. Shelton

Law and Artificial Intelligence
Woodrow Barfield and Ugo Pagallo

Politics of International Human Rights
David P. Forsythe

Community-based Conservation
Fikret Berkes

Global Production Networks
Neil M. Coe

Mental Health Law
Michael L. Perlin

Law and Literature
Peter Goodrich

Creative Industries
John Hartley

Global Administration Law
Sabino Cassese

Housing Studies
William A.V. Clark

Global Sports Law
Stephen F. Ross

Public Policy
B. Guy Peters

Empirical Legal Research
Herbert M. Kritzer

Cities
Peter J. Taylor

Law and Entrepreneurship
Shubha Ghosh

Mobilities
Mimi Sheller

Technology Policy
Albert N. Link and James Cunningham

Urban Transport Planning
Kevin J. Krizek and David A. King

Legal Reasoning
Larry Alexander and Emily Sherwin

Sustainable Competitive Advantage in Sales
Lawrence B. Chonko

Law and Development
Second Edition
Mariana Mota Prado and Michael J. Trebilcock

Law and Renewable Energy
Joel B. Eisen

Experience Economy
Jon Sundbo

Marxism and Human Geography
Kevin R. Cox

Maritime Law
Paul Todd

American Foreign Policy
Loch K. Johnson

Water Politics
Ken Conca

Business Ethics
John Hooker

Employee Engagement
Alan M. Saks and Jamie A. Gruman

Governance
Jon Pierre and B. Guy Peters

Demography
Wolfgang Lutz

Environmental Compliance and Enforcement
LeRoy C. Paddock

Migration Studies
Ronald Skeldon

Landmark Criminal Cases
George P. Fletcher

Comparative Legal Methods
Pier Giuseppe Monateri

U.S. Environmental Law
E. Donald Elliott and Daniel C. Esty

Gentrification
Chris Hamnett

Family Policy
Chiara Saraceno

Law and Psychology
Tom R. Tyler

Advertising
Patrick De Pelsmacker

New Institutional Economics
Claude Ménard and Mary M. Shirley

The Sociology of Sport
Eric Anderson and Rory Magrath

The Sociology of Peace Processes
John D. Brewer

Social Protection
James Midgley

Corporate Finance
James A. Brickley and Clifford W. Smith Jr

U.S. Federal Securities Law
Thomas Lee Hazen

Cybersecurity Law
David P. Fidler

The Sociology of Work
Amy S. Wharton

Marketing Strategy
George S. Day

Scenario Planning
Paul Schoemaker

Financial Inclusion
Robert Lensink, Calumn Hamilton and Charles Adjasi

Children's Rights
Wouter Vandenhole and Gamze Erdem Türkelli

Sustainable Careers
Jeffrey H. Greenhaus and Gerard A. Callanan

Business and Human Rights
Peter T. Muchlinski

Spatial Statistics
Daniel A. Griffith and Bin Li

The Sociology of the Self
Shanyang Zhao

Artificial Intelligence in Healthcare
Tom Davenport, John Glaser and Elizabeth Gardner

Central Banks and Monetary Policy
Jakob de Haan and Christiaan Pattipeilohy

Megaprojects
Nathalie Drouin and Rodney Turner

Social Capital
Karen S. Cook

Elections and Voting
Ian McAllister

Negotiation
Leigh Thompson and Cynthia S. Wang

Youth Studies
Howard Williamson and James E. Côté

Private Equity
Paul A. Gompers and Steven N. Kaplan

Digital Marketing
Utpal Dholakia

Water Economics and Policy
Ariel Dinar

Disaster Risk Reduction
Douglas Paton

Social Movements and Political Protests
Karl-Dieter Opp

Radical Innovation
Joe Tidd

Pricing Strategy and Analytics
Vithala R. Rao

Bounded Rationality
Clement A. Tisdell

International Food Law
Neal D. Fortin

International Conflict and Security Law
Second Edition
Nigel D. White

Entrepreneurial Finance
Second Edition
Hans Landström

US Civil Liberties
Susan N. Herman

Resilience
Fikret Berkes

Insurance Law
Robert H. Jerry, II

Applied Green Criminology
Rob White

Law and Religion
Frank S. Ravitch

Social Policy
Second Edition
Daniel Béland and Rianne Mahon

Substantive Criminal Law
Stephen J. Morse

Cross-Border Insolvency Law
Reinhard Bork

Behavioral Finance
H. Kent Baker, John R. Nofsinger, and Victor Ricciardi

Critical Global Development
Uma Kothari and Elise Klein

Private International Law and Procedure
Second Edition
Peter Hay

Victimology
Sandra Walklate

Party Politics
Richard S. Katz

Contract Law and Theory
Brian Bix

Environmental Impact Assessment
Second Edition
Angus Morrison-Saunders

Tourism Economics
David W. Marcouiller

Service Innovation
Faïz Gallouj, Faridah Djellal, and Camal Gallouj

U.S. Disability Law
Peter Blanck

US Data Privacy Law
Ari Ezra Waldman

Urban Segregation
Sako Musterd

Behavioral Law and Economics
Cass R. Sunstein

Economic Anthropology
Peter D. Little

ELGAR ADVANCED INTRODUCTIONS: LAW

www.advancedintros.com

Access the whole eBook collection at a cost effective price for law students at your institution.

Email: **sales@e-elgar.co.uk** for more information